REGIONAL VISIONARIES AND METROPOLITAN BOOSTERS

Decentralization, Regional Planning, and Parkways During the Interwar Years

REGIONAL VISIONARIES AND METROPOLITAN BOOSTERS

Decentralization, Regional Planning, and Parkways During the Interwar Years

by

Matthew Dalbey
Jackson State University, U.S.A.

KLUWER ACADEMIC PUBLISHERS
Boston / Dordrecht / London

Distributors for North, Central and South America:
Kluwer Academic Publishers
101 Philip Drive
Assinippi Park
Norwell, Massachusetts 02061 USA
Telephone (781) 871-6600
Fax (781) 681-9045
E-Mail <kluwer@wkap.com>

Distributors for all other countries:
Kluwer Academic Publishers Group
Post Office Box 322
3300 AH Dordrecht, THE NETHERLANDS
Telephone 31 786 576 000
Fax 31 786 576 254
E-Mail <services@wkap.nl>

Electronic Services <http://www.wkap.nl>

Library of Congress Cataloging-in-Publication Data

Dalbey, Matthew, 1965-
Regional visionaries and metropolitan boosters : decentralization, regional planning, and parkways during the interwar years / by Matthew Dalbey.
p. cm.
Includes bibliographical references and index.
ISBN 1-40207-104-3 (acid-free paper)
1. Regional planning—United States—History—20th century. 2. Parkways—United States—History—Case Studies. I. Title.

HT392.D35 2002
307.1'2'0973—dc21

2002067818

Printed on acid-free paper.
Printed in the United States of America

For Mollie, Noah, and Liz

Contents

Acknowledgements

Many individuals have helped me over the years it has taken to complete this work. I thrived on all the kind words and advice offered during this time.

I am grateful for the time and effort of the faculty in the Department of Urban Planning at Columbia University, including Elliott Sclar and Peter Marcuse, as well as Cliff Ellis at SUNY Albany. Two other colleagues, Randy Mason and Paul Sutter, provided constant support over the research and writing phases of this work. Their efforts – as motivators, critics, sounding boards, and peers – kept me going and helped turn this entire experience into a rewarding one. Thank you both.

I am also grateful to Jackson State University for financial and professional assistance with this project.

During the initial stages of research, I received a great deal of assistance from the archivist at the Shenandoah National Park Archives in Luray, Virginia and the librarians at the Vermont Historical Society in Montpelier, Vermont. Thanks also to the staff archivists at the National Archives Record Center in College Park, Maryland and the Virginia State Library in Richmond, Virginia.

Thanks also to Glenn Peake, Kelly Higgins, and Jan Hillegas for editing work on the various drafts. I also appreciate the help I received with the graphics from Richard E. Lloyd, Milo Dalbey, and Chris Dorin.

Although a number of people helped make this work better, any shortcomings are entirely my responsibility.

Finally, thanks to my spouse, Liz Hunter, for putting up with me during this seemingly never-ending process – your support and interest in my work helped more than you can ever know.

Preface

This book is an examination of two conflicting regional planning ideologies and the impact of this conflict on the development of two regional parkways. I hypothesize that regional parkways of the 1920s and 1930s emerged out of these two visions of regional planning – regionalism and metropolitanism. The regional view coalesced around the work of Benton MacKaye, Lewis Mumford, and the Regional Planning Association of America. The metropolitan viewpoint, while less definable, grew out of the market-oriented economic boosterism efforts associated with early twentieth century planning. This view found literal and philosophical support with Thomas Adams and the Regional Plan of New York and Its Environs. In an effort to flesh out the competing theories and the development of the regional parkway, I discuss the history of the Skyline Drive and the proposed Green Mountain Parkway.

In addition to supplementing the planning history and theory literature, I try to inform on issues important to the contemporary planning profession. The regional visionaries viewed their regional work as a social reform effort. The metropolitanists wanted to tweak the market so as to provide for a minimized congestion and economic hardship for the greatest number of citizens. This "vision versus reality" still troubles the profession today, especially in the areas of sustainable development, growth management, and "smart growth."

Matthew Dalbey
Jackson, Mississippi
March 2002

Chapter 1

Decentralization and Regional Planning

Practical and Ideological Problems

1. INTRODUCTION

Between 1921 and 1936 planners, landscape architects, business boosters, and automobile and tourism advocates planned for, built, and attempted to institutionalize regional parkways as major recreational and travel arteries for Americans. Prior to this period parkways existed within cities and early suburbs and had roots in early urban reform. Their use as facilitators of suburban development grew in the first part of the 20th century. The period between 1921 and 1936 saw the introduction of the automobile into the countryside and revealed a broader debate over development in the region. Two conflicting visions of regional development characterized this debate – called in this work the "regionalists'" vision and the "metropolitanists'" vision. The debate continues today and is expressed in two ways. One is the debate over metropolitan sprawl, loss of farmland, consumerism, and what constitutes appropriate road planning. The second revolves around the very core of the planning profession: Should planners work to present visions of future social and economic development or should their work accommodate the market and tweak it only when seriously needed?

The purpose of this study is to explore in detail how the two planning ideologies shaped the parkway in the region. Previous histories have focused on design, use, and, to some extent, the institutions and ideology. These histories leave out the social consequences and portray the regional parkway as a benign intervention into vacant landscapes. The regionalists and the metropolitanists made attempts to structure the decentralization of congested urban areas. The conflict between the regionalists and the metropolitanists shaped the form and meaning of the regional parkway during the 1920s and 1930s. This is the story of that conflict.

1.1 The Regionalists

"Regionalists" describes a distinct group of regional planners – really regional visionaries – primarily associated with ideas that evolved out of the work of the Regional Planning Association of America (RPAA). They believed that regional planning in a capitalist democracy required the vision to look beyond the conventional logic of the city and the market. This vision required planning tools specific to the region and its component parts, as well as the cultural and political will to fundamentally break from the dominant economic forces. Unfortunately the regionalists' vision was not always the practical path towards reform but it did present an alternative to the path facilitated by the metropolitanist planners.

1.2 The Metropolitanists

"Metropolitanists" attempted to plan regionally by facilitating the expansion of the market. They employed the tools available to city planners on a regional scale and attempted to accommodate (rather than fundamentally reform, as the regionalists did) the inefficiencies associated with regional urbanization (or metropolitanization). The metropolitanists were business boosters, developers, and planners interested in tweaking the market so as to interfere only minimally with its logic.

2. PREMISES FOR THE STUDY OF THE REGIONAL PARKWAY

During the years between World War I and World War II automobile travel on the regional parkways (and later freeways) in the United States supplanted rail and foot travel as the dominant mode of access to the regional landscape. The builders and supporters of the regional parkway (that is, the metropolitanists) included planners, landscape architects, automobile enthusiasts, business interests, some environmentalists and conservationists, and many local, state, and federal officials. The regional visionaries, on the other hand, viewed the introduction of the automobile as antithetical to the pursuit of regional culture. Given the terms of the debate, the regional parkway left its mark on indigenous communities, the conceptualization of the natural environment, regional planning theory and practice, and the landscape itself. The history of the regional parkway indicates that early parkways, such as Skyline Drive in Virginia, evolved out of a closed planning process without public input. Later parkways, like the proposed Green Mountain Parkway in Vermont, had a more open, institutionalized structure, but supporters of the parkway could not gain the political support for its implementation. Many of the issues that confronted proponents of the regional parkway came back during the 1950s and 1960s in the form of controversies over the construction of the interstate highway system. Moreover, contemporary planning issues such as sustainable resource use, conservation, smart growth policies, and environmental protection on the one hand, and property rights, deregulation, and subsidized sprawl on the other are the updated form of the conflict that shaped the regional parkway between 1921 and 1936.

This study demonstrates that even during an era when regional planning had a national constituency, problems with implementation still existed. In order to create a successful initiative, visionary planners had to try to develop a constituency of supporters who believed in reformist goals. Benton MacKaye, one of the leading regionalists, believed that the construction of the Appalachian Trail by middle-class professionals would lead to a "social readjustment," with the same professionals leading the charge. Instead, these middle-class volunteers put completion of the Trail ahead of reformist goals (most had previously rejected the "social readjustment" goals anyway) and a segment of these volunteers saw no need to hold out against the parkways to save the primacy of the Trail on the ridgeline.[1]

The history of the regional parkway and its evolution from the conflict between the regionalists and the metropolitanists can provide us with practical knowledge about the current state of the planning profession. It suggests that neglecting a progressive vision for the comfort of market facilitation is too easy and often dangerous, because too many people are left out of the market. Much of what we value as a society is similarly left out. This study presents to planning professionals a picture of progressive planners as visionaries or leaders in a field often filled with market facilitators. Moreover, by restating and broadening the 1932 debate between Lewis Mumford and Thomas Adams over the Regional Plan of New York and its Environs, this work re-examines one of the essential ongoing debates in regional planning.

2.1 The Questions

Two conflicting ideologies shaped the development of regional parkways in the United States between 1921 and 1936. Adherents to these two ideologies attempted to rationalize a number of contradictions in the period between World War I and World War II. These contradictions included:

1. City planning tools versus the usefulness of these tools when planning for a variety of values that are not necessarily urban (that is to say, economic, cultural, social reform and progressive ideas versus the dominant economic paradigm).
2. The small town versus the sameness of urbanization, the region versus the metropolis.
3. The indigenous versus the metropolitan.

There existed no guiding doctrine for planning the parkway in the region prior to the early part of the twentieth century. In fact, neither did regional planning have any adequate model prior to this period. Rather, the competing regional planning groups – the regionalists and the metropolitanists – worked to gain dominance for their respective theories. In a very broad way, the regionalists and the metropolitanists each had roots in the Garden City movement in England and other European utopian planning efforts. By the end of World War I, the metropolitanists were ready to employ the tools of city planning in regional planning. They believed that rational and efficient planning could occur in the region using the same tools used for the rational and efficient city. The regionalists believed that regional planning required an entirely different set of tools, and that city planning in the region was anathema to the vitality of the region. Moreover, the future of the city – decentralization necessary to relieve congestion – depended upon the regionalists' views.

The proponents of the regional parkway had limited precedents. The regional parkway, according to contemporary supporters, could have served several purposes. Proponents believed it could serve as a connector road between two places, as access for a population to recreational opportunities, as infrastructure to help increase land values of the adjacent property owners. The regional parkway may have been modeled on roads and highways used as rural and regional development tools, or even conceived as the more technologically innovative form of Benton MacKaye's Appalachian Trail proposal.[2] Whereas the urban freeway emerged as a hybrid of the different visions of highway development, according to Cliff Ellis,[3] the regional parkway evolved from the conflict over conceptions of the region and the competing visions associated with those concepts.

Methodologically, this is an historical study. The Green Mountain Parkway and Skyline Drive initiatives revealed and dramatized the evolving planning issues and demonstrated the increasing institutionalization of the planning process. The research shows that even though the two parkway initiatives began only years apart, they allowed for different amounts of public participation and demonstrated to varying degrees the ability of the extant communities to get involved.[4]

There are three basic parts to the research. The first is an exploration of the roots of the regional parkway, from the cemetery road to the park road and then to the suburban parkway. The early history of the regional parkway is drawn from a thorough review of both primary and secondary texts and required the examination and reexamination of the professions involved, including planners, engineers, architects and landscape architects, as well as the institutional actors. This analysis of the early parkways demonstrates how the professionals viewed their own historical models and used that understanding in their own work.

Important works discussing the early history of the park and parkway vary in their usefulness. Christian Zapatka wrote two pieces on the parkway in the United States. The first, an article in a 1987 issue of *Lotus International* entitled "The American Parkways: Origins and Evolution of the Park-Road," is narrowly focused on the parkway as design artifact. The second, a monograph entitled *The American Landscape*, is similarly concerned with design, yet places the development of the parkway within the broader context of the development of the American built environment.[5]

Other works discussing the early history of the park and parkway similarly present the development of the park-road within the context of other histories. David Schuyler, in his work *The New Urban Landscape*, discusses the early roots of the park movement in 19th century American cities as parkways and park roads evolved out of reform efforts associated with the industrializing cities.[6] In *Design on the Land*, Norman Newton discusses the development of the parkway as part of his design history of the American landscape.[7] And Ethan Carr, in his book *Wilderness by Design*, analyzes the regional parkway as a design artifact.[8]

The second task is to outline the two conflicting theories of regional planning. The sources for this task were varied and included many primary and secondary sources. The dominant ideology associated with planning the regionalists' parkway was that of the RPAA. Benton MacKaye's article entitled "An Appalachian Trail: A Study in Regional Planning"[9] was part of the foundation, along with the Garden City movement in England and the progressive social and economic ideology of the RPAA and many of its collective and individual concepts of regional planning. His later writings on parkways (both in support and opposition) were rooted in and consistent with the Appalachian Trail concept.

The metropolitanists' vision for the region grew out of traditional city planning models, with roots in the City Beautiful, the City Efficient, and the Garden City. Business boosters, road associations, the supporters of the Regional Plan of New York and Its Environs (RPNY & E), and planners such as Thomas Adams and John Nolen represented this metropolitan ideology.

The sources for this section include the MacKaye Family Papers and primary and secondary texts on the RPAA, the works of Lewis Mumford and Thomas Adams, and primary sources from the RPNY & E, the Bronx River Parkway, and the Westchester County Parkways.

Benton MacKaye published a number of texts and articles on his concept of regionalism and regional planning. *The New Exploration: A Philosophy of Regional Planning*, first published in 1928, and *From Geography to Geotechnics*, published in 1969, and a series of articles in *The Survey Graphic* are the most important.[10] Lewis Mumford's works, including *The Story of Utopias*, *The Golden Day*, and numerous articles published in journals and magazines, gave important insights on the regionalists' vision.[11]

Secondary sources associated with the regional vision include Carl Sussman's work, *Planning the Fourth Migration: The Neglected Vision of the Regional Planning Association of America*, Mark Luccarelli's *Lewis Mumford and the Ecological Region*, and Edward Spann's *Designing Modern America: The Regional Planning Association of America and its Members*.[12] Spann's well-researched efforts provide the best historical analysis of the RPAA.

The metropolitanists' literature is more narrowly defined and includes the writings of Thomas Adams, his work for *The Regional Plan of New York and Its Environs*, *The Regional Survey of New York and Its Environs*,[13] as well as periodically published conference proceedings. Other primary sources include John Nolen and Henry Hubbard's *Parkways and Land Values*, Stanley Abbott's writings on Westchester, and work by Gilmore Clarke, among others.[14] As secondary sources, David Johnson's *Planning the Great Metropolis* and Robert Caro's *The Power Broker* provide a critical baseline for the study of the metropolitanist ideology.[15]

The case studies are the third task. The Skyline Drive was the earliest of the regional parkways planned in the eastern United States. Although proposed to follow the lead of the Skyline Drive, the citizens of Vermont defeated a referendum that would have authorized the construction of the Green Mountain Parkway in 1936. The history of these two parkway initiatives is the chronicle of the regional parkway in the East, from its infancy as a local and state-level initiative brought about by the regional elite to the more democratic, open planning process initiated at the federal level. The two parkways were intended to make the wilderness (or at least a conception of the wilderness) accessible to the general public, bring about economic opportunity, and help to establish a greater constituency for the National Park Service in the eastern part of the United States. Regionalist planners such as Benton MacKaye supported wilderness accessibility; however, he grew increasingly pessimistic about the ability of the wilderness to survive the assault by the automobile. In the case of the Skyline Drive, the Potomac Appalachian Trail Club (PATC) was in its infancy and had only limited political strength with which to contest the Drive. Moreover, the PATC leadership was ultimately not interested in continually contesting the Drive as built. On the other hand, the Green Mountain Club in Vermont (which worked against the development of the Green Mountain Parkway) had a longer history as a club, having built The Long Trail during the decade prior to MacKaye's publication of the Appalachian Trail idea.

Had the Green Mountain Parkway been built as planned, its impact on the regional landscape would likely have been much the same as that of the Skyline Drive. Contemporary supporters of the Green Mountain Parkway looked to the Skyline Drive effort to support their case. A Burlington, Vermont, newspaper editor surveyed a number of central Virginia newspapers and civic groups in an effort to help the cause in Vermont. The response from Virginia was generally positive, yet there was a general bafflement among the Virginians over the idea that local involvement could have had any influence on the Skyline Drive effort.[16]

The Skyline Drive has a significant and mostly unpublished historical record. Moreover, the history of the Skyline Drive is intimately tied to the history of the Shenandoah National Park. For the most part, the Shenandoah National Park papers and records are available through the Park archive in Luray, Virginia. Additional papers are available at the Park Service archive in Harpers Ferry, WV, and at the National Archives Records Center in College Park, MD. Other documents associated with the early history of the Skyline Drive are available at the Virginia State Library. Other sources include local and statewide newspapers, Chamber of Commerce bulletins, and newsletters of the various civic clubs such as the Potomac Appalachian Trail Club. Secondary source material on the Shenandoah National Park and Skyline Drive is limited.

Study of the proposed Green Mountain Parkway presented many of the same methodological challenges as the Skyline Drive. The primary studies for the parkway were those conducted by landscape architects associated with the National Park Service. Their analyses are available at the Vermont Historical Society and in the special collections section of the University of Vermont library. The National Park Service Archives in Harpers Ferry contains limited additional material, as does the National Archives Records Center. Materials regarding supporters and opponents of the parkway are available at the Vermont Historical Society, the University of Vermont, and in local and statewide newspapers. In addition, the Vermont State Chamber of Commerce published many bulletins referencing the parkway, while one of the main parkway opponents, the Green Mountain Club, published its own bulletins. As with resources for study of the Skyline Drive, secondary source material regarding the Green Mountain Parkway is limited.

3. ORGANIZATION OF THE STUDY

This study consists of seven chapters. Chapter I has introduced the study, the major themes and considerations. Chapter II discusses the prehistory of the regional parkway and its roots within the 19th century urban reform movement. Chapters III and IV outline the two conflicting views of regional development – the regional vision and the metropolitan reality. Chapter V is a case study of the Skyline Drive. Chapter VI is the Green Mountain Parkway case study. Chapter VII is the conclusion and contains an analysis of the broader connections associated with this study.

---

[1] See Ronald Foresta, "Transformation of the Appalachian Trail," *Geographical Review* 77 (1987): 76-85.

[2] In "Visions of Urban Freeways, 1930-1970," diss., U California Berkeley, 1990, 16, Cliff Ellis indicates a similar idea with regard to urban freeways. He writes, "as guides to policy, [the different visions of urban freeways] lead to different urban patterns. These images were often seen in hybrid form, blended together to suit shifting economic, political, and professional currents."

[3] Ellis, "Visions of Urban Freeways,"16-17.

[4] In "On the Meaning and Analysis of Change in Social History," *Theory, Method, and Practice in Social and Cultural History*, eds. Peter Karsten and John Modell (New York, London: NYU P, 1992) 33-56.

[5] Christian Zapatka, "The American Parkways," *Lotus International* 5 (1987): 96-128, and Zapatka, *The American Landscape*, ed. Mirko Zardini (New York: Princeton Architectural P, 1995).

[6] David Schuyler, *The New Urban Landscape: The Redefinition of City Form in Nineteenth-Century America* (Baltimore and London: Johns Hopkins U P, 1986).

[7] Norman T. Newton, *Design on the Land: The Development of Landscape Architecture* (Cambridge, MA and London, England: The Belknap P of Harvard U, 1971).

[8] Ethan Carr, *Wilderness by Design: Landscape Architecture and the National Park Service* (Lincoln and London: U of Nebraska P, 1998).

[9] Benton MacKaye, "An Appalachian Trail: A project in regional planning," *Journal of the American Institute of Architects* 9 (1921): 325-330.

[10] Benton MacKaye, *The New Exploration* (1928; Urbana: U Illinois P, 1962), and MacKaye, *From Geography to Geotechnics*, ed. Paul T. Bryant (Urbana: U of Illinois P, 1968).

[11] Lewis Mumford, *The Story of Utopias* (New York: Boni and Liveright, 1922), and Mumford, *The Golden Day: A Study in American Experience and Culture* (New York: Boni and Liveright, 1926).

[12] Carl Sussman, *Planning the Fourth Migration: The Neglected Vision of the Regional Planning Association of America* (Cambridge: MIT P, 1976), Mark Luccarelli, *Lewis Mumford and the Ecological Region: The Politics of Planning* (New York and London: The Guilford P), Edward K. Spann, *Designing Modern America: The Regional Planning Association of America and Its Members* (Columbus: Ohio State U P, 1996).

[13] Committee on the Regional Plan of New York and Its Environs, *The Graphic Regional Plan*, Vol. 1 (New York: Committee on the Regional Plan of New York and Its Environs, 1929), and Committee on the Regional Plan, *Regional Survey of New York and Its Environs* (New York: Regional Plan of New York and its Environs, 1928).

[14] John Nolen and Henry V. Hubbard, *Harvard City Planning Studies, Volume XI: Parkways and Land Values* (London: Oxford U P, 1937).

[15] David A. Johnson, *Planning the Great Metropolis: The 1929 Regional Plan of New York and Its Environs* (London: E & F N Spon, 1996), and Robert Caro, *The Power Broker: Robert Moses and the Fall of New York* (New York: Vintage Books, 1974).

[16] See, for example, R.S. Fansler to David W. Howe, 14 January, 1936, Doc T13, James P. Taylor Papers, Vermont Historical Society, Montpelier, VT.

Chapter 2

From Urban Reform to Parkway

The Early History of the Urban Park Road and the Parkway

It is doubtful that any single type of park area has been more widely misunderstood and misinterpreted than the parkway. The confusion is hardly to be wondered at when one considers with what free and easy imprecision the term "parkway" has been used.[1]

Norman Newton, 1971.

1. INTRODUCTION

The early history of the parkway in the United States demonstrates that early city planners and landscape architects designed the urban and suburban parkway to serve the city – as a method of urban reform. By showing that the parkway has its roots in the city, this chapter lays the groundwork for the later conflict between the regionalist and metropolitanist planners as they try to put the parkway into the region.

If we are to understand the conflict over the planning of the regional parkway, it is important to understand that its early history indicates that the urban park road and the suburban parkway strongly influenced the conceptualization of the parkway in the region. Planners and supporters of regional parkways looked back to urban and suburban parkways as a guide for their own endeavors.

In his article "The American Parkways," Christian Zapatka summarizes Charles Eliot's description of parkways by "likening parkways to parks" and further defining a parkway as "essentially a road running through a park."[2] This definition fits the design-oriented view that the parkway in the United States was on a historical trajectory that brought it out from the city, to the suburb, and then to the countryside. Zapatka uses the terms "the urban, the suburban and the national."[3] This design-oriented description suggests a lack of controversy and thus fails to account for the conflict over planning in the region.

In a sense, the history of early parkways and park roads is the history of urban reform and city planning, which city planning and urban historians have noted.[4] Unfortunately, the remainder of the history of parkways has largely been left to historians of design artifacts, who seem to forget the regional parkway's roots in reform and the early city planning movement.[5]

The parkway is different from the highway. The original "parkway" integrated the natural reserve – the park itself – within the urban fabric. Zapatka's analysis, for instance, views the "national" or regional parkway as the foundation for the "road running through nature" and reconstructs the history to point in this direction. In actuality, urban reform produced the urban park and parkway. The suburban parkway served the city as it grew and expanded, while the regional parkway developed out of the conflict over regional reform – reform articulated by the debate between the regionalists and the metropolitanists.

2. PARKS AND PARKWAYS AS URBAN REFORM

2.1 The Rural Cemetery Roots

The rise of industrialization and technology in the nineteenth century brought about changes in the urban form of the United States. The history of this change is well documented and broad.[6] In *The New Urban Landscape*, David Schuyler argues that nineteenth century urban form evolved in three significant ways. First, the urban fabric opened up. Prior to industrialization, the economics of life required compact spaces. Industrialization and technology allowed for decentralization. Second, the growing middle-class desire to preserve and increase property values encouraged the development of public amenities. And third, technological advances, especially in transportation, allowed for the separation of space by use. People no longer needed to live where they worked.[7] These changes led to a new urban form and provided ample opportunity for the creation of parks and parkways. The justification for this public infrastructure – the introduction of parks into the urban fabric – often dictated the discourse of reform.

While the broad philosophical impetus for parks and parkways rested in urban reform, it was the early rural cemetery that served as the physical model for reform. The first "rural" cemetery, Mount Auburn Cemetery in

Cambridge, Massachusetts, was conceived by Dr. Jacob Bigelow, a Boston physician aware of the health consequences of overcrowded graveyards in cities and interested in establishing an appropriate place for burials.[8]

The term "rural" is a bit misleading and needs some explanation. According to historian Thomas Bender, "rural cemetery" meant an appropriately landscaped ground at the edge of the city.[9] David Schuyler further points out that the burial ground outside the city had classical antecedents that were both practical and spiritual.[10] The concept of separation fits the modern industrialized city. However, as a place for recreation and leisure for the living, the rural cemetery blurred these lines of separation.

Mount Auburn opened in 1831 with the express purpose of serving the living, not the dead. John W. Reps notes in *The Making of Urban America* that many of the cemetery's visitors used the space as a pleasure ground.[11] By the late 1840s, horticulturist and rural landscape architect Andrew Jackson Downing stated that nearly 30,000 people visited Mt. Auburn in a single year. Similar numbers visited Laurel Hill Cemetery in Philadelphia and Greenwood Cemetery in Brooklyn.[12] The visitors did not disturb the "sponsors," as Reps commented, but they needed some sort of public recreation ground.

In October 1848, Downing, writing in the *Horticulturist*, commented on the use of cemeteries as pleasure grounds – with a view toward the development of parks.

> Indeed, these cemeteries [Mt. Auburn, Greenwood, and Laurel Hill] are the only places in the country that can give an untravelled American any idea of the beauty of many of the public parks and gardens abroad. Judging from the crowds of people in carriages, and on foot, which I find constantly thronging Greenwood and Mt. Auburn, I think it is plain enough how much our citizens, of all classes, would enjoy public parks on a similar scale.[13]

Downing saw public parks as the epitome of a republican form of government, noting that even France and Germany, previously thought of as anything but republican, had set aside land reserved for enjoyment by all classes.[14] The movement to build rural cemeteries in close proximity to the industrializing cities served as a prelude and model for the building of parks in American cities.

David Schuyler points out three ways the cemetery experience influenced park development and urban reform. First, cemeteries encouraged individuals to pursue self-reflection and realize the value of the natural world for self-healing purposes. Second, the cemetery served as a place that could inspire thoughtfulness and awe, hence Schuyler's use of the term "didactic landscape" to describe the rural cemetery. Finally, the cemetery provided for an existing and accessible alternative to the city's congestion.[15]

2.2 The Early Public Parks and Parkways in the Nineteenth Century

The early parks in the United States, beginning with Central Park in New York City, have a rich historiography. The historical literature builds upon two primary themes: the park as design artifact and the park as social artifact. The historiography of the urban park and parkway as design artifact minimizes the roots in urban reform and distorts the later history of the regional-scale parkway. On the other hand, the park and parkway seen as a social artifact allows for the best understanding of the later history of the regional parkway.

2.2.1 Central Park

Many historians trace the park movement directly to the reform and aesthetic ideas associated with the rural cemetery as well as to the landscape gardening work of A.J. Downing. As the first large-scale public park based upon the aesthetic and reform doctrines derived from the rural cemeteries, Central Park emerged as the primary model for the public park in the United States.[16]

This direct, neatly defined lineage from the aesthetic and reform notions of Downing, park advocate and newspaper editor William Cullen Bryant, and their supporters does not tell the entire story. Historians Roy Rosenzweig and Elizabeth Blackmar point out in their work, *The Park and the People: A History of Central Park*, that property speculation, commercial interests, political patronage, and job provision entered into the initial motivations for Central Park.[17]

Frederick Law Olmsted (1822-1903) and Calvert Vaux (1824-1895) won the competition for the commission to build Central Park. Olmsted became superintendent in charge of the labor force at the yet-to-be-built park and under the supervision of Chief Engineer Egbert Viele, began to prepare the grounds for the development of the park. Up to this point in his career, Olmsted had traveled and written extensively on farming and traveling, and as a journalist had written a series on the South and slavery for the *New York Times*. Vaux, who had emigrated from England to the United States in 1850 to work with A.J. Downing, talked Olmsted into submitting a plan to the 1858 competition sponsored by the Board of Commissioners of the Central Park. On April 28, 1858, the Board of Commissioners chose the Olmsted and Vaux plan – Plan No. 33, entitled "Greensward" – as the competition winner. In May 1858, the Board of Commissioners appointed Olmsted "Architect in Chief of the Central Park."[18]

Olmsted and Vaux designed a plan for a park that would immediately let visitors know that they had left the city and entered, in their view, a natural landscape. Many of the others in the competition had placed architectonic monuments at the center of the park plan. Olmsted and Vaux purposely avoided this, as did others who believed in the writings and work of A.J. Downing and European-influenced park design.[19]

Francesco Dal Co, in his essay on the ideology of the park movement in the United States, points to the importance of Central Park and the work of Olmsted and Vaux. The designers basically took the Downing concept of the park setting and inserted it into the urban fabric.[20] The park plan provided for a break from the urban by offering up to the visitor a tamed and picturesque nature. Olmsted and Vaux sought to reform the urban by bringing within the urban an idealized version of the anti-urban.

The pedestrian, the carriage rider, and the horseback rider each had grade-separated carriageways and paths. The carriageways allowed for controlled scenic experiences, upon which Olmsted and Vaux built their vision of the urban park. Although Olmsted and Vaux planned and built the park within an urban environment, the design of the carriageways made use of the natural and constructed settings to the visitor's advantage.[21] The 59th Street entrance quickly took the visitor to the center of the Park – immediately immersing her or him in the pastoral landscape.[22] Similarly, the separated grade crossings and the sinking of the three cross-streets that transect the Park allowed for the aesthetic – such as the privileging of the park over commerce – to prevail in the experience.

Olmsted and Vaux separated the transverse streets of the pre-existing city grid from the carriageways and the rest of the park by limiting access and separating the grade of the streets. This concept, while rooted in and viewed as design, is more broadly a social and reform concept. In recognizing the desire to bring the natural landscape into the city, Olmsted and Vaux separated out the city of commerce while not allowing the park to interfere with the day-to-day workings of that city.[23] The path system allowed visitors in carriages to circulate throughout while not interfering with the enjoyment of others. In his book, *The Park and the Town: Public Landscape in the 19th and 20th Centuries*, George F. Chadwick argues that the circulation system was designed in anticipation of further congestion and provided a model for towns that later were required to deal with the influx of automobile traffic.[24] Carriageways, the grade-separated intersections, and the ability of Central Park to function as a park in a developing urban environment were testament to the forward-looking work of Olmsted and

Vaux. These design innovations were successfully modeled and replicated in subsequent urban and regional parks and parkways.

In the end, "Greensward " and the ultimate completion of Central Park had many important aspects and influences. First, landscape and scenery became part of the daily lives of city dwellers. Property values of land adjacent to the park rose and the nearby park provided the healthful benefits of light and air to New Yorkers. Central Park thus helped to mitigate unhealthiness associated with an overcrowded nineteenth century New York City. And, finally, Central Park allowed New York City to take its place among other cities of the world where public parks had been built.[25]

Central Park has influenced many planners, landscape architects, and other practitioners since 1858. For the purpose of this study, Central Park serves as the first example of a publicly constructed and publicly financed natural ground created explicitly for public use. Further, the grounds were designed for active recreation – walking, riding horses, and riding in carriages. The carriageways at Central Park and the grade separations at the intersections of the roads/paths used by each mode of transportation served as models for the new parkways subsequently constructed in the United States. European models[26] – specifically neo-classical models – influenced the rhetoric of future parks and parkways, as well as the landscape.

This view of the initial park movement and Central Park as urban reform should not be lost. Proponents of Central Park supported its construction for numerous reasons. The change parks brought to the urban landscape was abrupt and indeed a method of reform in and of itself as the urban park provided an alternative to the industrializing city of the late nineteenth century. Schuyler notes that the reform movement associated with the introduction of the natural landscape into the city changed as the nineteenth century ended. By 1900, the City Beautiful had brought about the concept of the monumental, classical city, which replaced the more romantic vision of an urban reform movement through parks.[27]

In the current view, which was developed over the past 100 years or so, reform associated with the parks movement obscured many of the broad urban problems – poverty, housing shortages, crime, sanitary concerns, and the drain of suburbanization, among others. The push for urban parks often limited the scope of other reform efforts. However, the urban park movement begun with Central Park, and, viewed within the context of the mid-nineteenth century city, had reform at its core.[28] Arguably, the urban park movement was truly democratic at its base. The valid critique, however, is that the introduction of parks as the primary component of reform was far from comprehensive.

Central Park existed as the model for planners from the outset. As an artifact in the broader discussion of the regional parkway, its roadways, carriageways, and grade separations served as models for landscape architects and planners over the next century. In a broader sense, Central Park served as a model for both the "Regional Visionaries" and the "Metropolitanists" alike. Central Park demonstrated that planners, landscape architects, and social reformers could move from vision to actuality and lead through planning. Central Park also demonstrated that its reformist planning accommodated market interests and likely only "tweaked" the market while providing for increased property values and the better circulation of goods and commerce.

2.2.2 Prospect Park

Olmsted and Vaux (primarily Vaux because of Olmsted's absence in California) finished the preliminary planning of Prospect Park in New York in 1866. Many of the ideas first presented in Central Park were refined in Prospect Park – grade-separated roads and paths, varieties in the scenery, and the relationship between the park and the surrounding neighborhoods, for instance. Indeed, historians and, some believe, Olmsted and Vaux themselves, viewed Prospect Park as superior to Central Park.[29] Most important to this study is the maturation of the idea of the parkway. Olmsted and Vaux expended a good amount of effort conceptualizing the parkway within the context of the growing urban fabric as well as linking it to European precedent.[30] This conceptualization is reflected in the landscape itself.

The New York State Legislature set aside the land for Prospect Park in three separate acts in 1861, 1866, and 1868. The final act provided an area that allowed the Olmsted and Vaux design to incorporate the largest section of open space and a good portion of the carriage road that circumscribed the entire park. Landscape historian Norman Newton points out that although the first two acts of the New York State Legislature had not included this last section, Vaux and Olmsted had always advocated for additional land in their earliest collaborative reports to the Board of Commissioners of Prospect Park.[31]

An analysis of the plan of Prospect Park demonstrates the initial concepts of the park designers. First, the circuit drive – the carriageway that circles the entire park – allowed the visitor to experience the three primary elements essential to the 526-acre park: turf, wood, and water.[32] Further, the grade separations incorporated into the circulation plan allowed for the construction of arches, portals, and tunnels that demarcated the different sectors of the park.

The site in Brooklyn provided a more freely formed design than the more restricted Central Park. At Prospect Park, Olmsted and Vaux went to great lengths to establish a more natural landscape within the growing urban environment. By constructing natural hillsides or shoulders (which separated the park from the city) along the outside edges of Prospect Park, the designers allowed the visitor to have an even greater natural experience.[33] More than Central Park, Prospect Park brought the natural landscape into the everyday life of the urban population.

Olmsted and Vaux solicited the Board of Commissioners of Prospect Park in a number of reports that touted the benefits of the park to the citizens of Brooklyn. In the January 1, 1868, report published in *The Papers of Frederick Law Olmsted* as "The Concept of the 'Park-Way,'" Olmsted and Vaux commented on the benefits of park planning in Brooklyn, as well as the broader vision of parks and parkways in growing towns. Citing the public use and enthusiasm for Prospect Park well before completion, the designers wrote of how the city's residents were beginning to make daily (or at least consistent) visits to the Prospect Park. They predicted that over time, as the work was completed, use would just continue to increase.[34]

Vaux and Olmsted continued championing the park idea and the benefits of parks within growing urban environments. In espousing these concepts, they took a decidedly historical view, noting that the growth in urban environments no longer posed health and safety concerns to the extent that they once did. They cited increases in urban population and advancements in civilization, "due mainly to the increase of facilities for communication, transportation and exchange throughout the world [. . .]."[35] The issues and problems, which had previously caused the "evils of large towns," included the narrowness of streets built prior to industrialization and the lack of open space and air available to the residents of cities.[36]

In an argument aimed more broadly at the Board of Commissioners of Prospect Park, Olmsted and Vaux advocated a number of concepts to promote the healthful growth of cities. In addition to the construction of parks – such as Prospect or Central Park – they called for a system of public transportation, noting the growing separation of domestic life and business life. Suburbs and the distance from the city's core limited the ability of people to "stroll out into the country in search of fresh air, quietness, and recreation."[37] New transportation and changes in the domestic-business relationship required opportunities for recreation, both active and passive.

In sum, the designers proposed the concept of the 'Park-Way.' They planned and proposed to connect the urban park with the other parts of the city, as well as with the suburbs, by employing a series of parkways. The parkways incorporated efficient transportation with the added benefit of a recreational area – tree-lined streets with walking areas and lanes for local traffic, separated by curbs – for walking or riding in a natural setting on the way to the main park. Olmsted and Vaux modeled their proposal on two primary inspirations, *Unter den Linden* in Berlin and the *Avenue de l'Imperatrice* in Paris. Each existed as a formal boulevard that connected the central part of the urban area with a pleasure ground or reserve adjacent to or within the urban area. These boulevards met, as Olmsted and Vaux wrote, "the several requirements which we have thus far examined, giving access for the purposes of ordinary traffic to all the houses that front upon it, offering a special road for driving and riding without turning commercial vehicles from the right of way, and furnishing ample public walks, with room for seats and with borders of turf in which trees may grow of the most stately character."[38]

The parkway concept evolved into the Eastern (1870-74) and Ocean (1874-76) Parkways in Brooklyn. While Olmsted and Vaux proposed an even greater network of smaller parks and parkways throughout Brooklyn and even a connector into Manhattan's Central Park, the lack of sufficient funds and the disinterest in the proposal on the part of the Board of Commissioners of Central Park ended the prospect of a more grand plan.[39] Instead, the Eastern and Ocean Parkways provided residents of Brooklyn with a sliver of Prospect Park running through neighborhoods. The parkways connected the neighborhoods to the natural preserve of Prospect Park while also facilitating traditional urban commercial uses.

While the designers did not completely implement this concept, they had proposed a new model for integrating parks with transportation.[40] As if in response to this new model, the City of New York, in 1888, acquired 4000 acres of parkland in the Bronx. Connections to this land were made through a series of "parkways" modeled on the Eastern and Ocean Parkway projects. The City of New York designed and constructed the Mosholu, Pelham, Bronx, and Croton Parkways following the ideas modeled on Brooklyn.[41]

2.2.3 Other Park Projects

While working on the Prospect Park plan as well as the plans for the connecting parkways, Olmsted and Vaux also began a park and parkway system in Buffalo, New York. The Olmsted and Vaux firm had a thirty-year relationship with the City of Buffalo. Their work between 1868 and 1898 included plans for three large-scale urban parks and a system of parkways that linked the parks and the city. Olmsted, according to Schuyler, believed that the work done by his firm and laid out over a pre-existing city plan provided Buffalo with the preeminent plan in the United States.[42]

In Buffalo, Olmsted had the opportunity to work with city officials before they had chosen the site for the parks. This was the first time in Olmsted's career that he had such an opportunity.[43] Fortuitously, Olmsted's involvement from this early point allowed him to create a more systematic and cohesive vision, "not of a single park but of a comprehensive system of parks and parkways."[44]

In the plan for the park and parkway system in Buffalo, the parkways linked the three parks and knitted them into the urban fabric. The first part of Olmsted's work, between 1868 and 1876, included three parks and a complementary parkway system, in collaboration with Calvert Vaux. The second part of this work, completed in 1898, was done by Olmsted, his son Frederick Law Olmsted Jr., and his nephew and adopted son, John Charles Olmsted.[45]

Designed and completed between 1868 and 1876 were 350 acres called simply "the Park" (now Delaware Park). Olmsted and Vaux designed the Park to take advantage of its pastoral scenery. As with their other projects, the natural setting provided a steady contrast to the urban fabric. In dealing with park traffic, Olmsted separated pedestrians from vehicle riders and presented, as much as possible, scenic views of the landscape without interrupting traffic flow. Again, as with earlier parks, Olmsted handled intersecting traffic in much the same way. He sank the crossing roads below the level of the park, hiding the roads' intersections from park users. Where the landscape did not allow the grade separations naturally, he employed bridges as well as shrub and tree plantings.[46]

In addition to the Park, Olmsted and Vaux designed in Buffalo two other smaller parks – "the Parade" (now called Martin Luther King Jr. Park) and "the Front" (along the Erie Canal). Parkways linked these two smaller parks with the Park as well as with the rest of the city. The parkways provided a green setting for the rest of the city residents – especially those who did not live adjacent to the Park, the Parade, or the Front.[47] Olmsted and Vaux wove nature into the urban fabric through the integrated use of parks and parkways.

The second stage of park development, completed by 1898, continued the earlier work and notably included smaller parks and a parkway system that accommodated the projected growth in the Buffalo metropolitan area.[48]

2.2.4 The Olmsted Legacy, Charles Eliot and the Boston Metropolitan Parks

Olmsted's work in Boston began in 1870 when he presented his paper, "Public Parks and the Enlargement of Towns," at the Lowell Institute. Anticipating population growth in Boston, Olmsted recommended park planning that recognized this inevitability and called for a broader vision for park development. Prior to Olmsted's presentation, H.W.S. Cleveland, a landscape designer who had worked with Olmsted and Vaux at Prospect Park, suggested, along with others, that the City of Boston develop a rational metropolitan park system. Lacking the public will to purchase land outside Boston, these ideas were never realized.[49]

In April of 1876 Olmsted recommended to the Boston Park Commissioners a system of four parks, at Charles River, Back Bay, Jamaica Pond, and West Roxbury. Acting upon Olmsted's suggestion, the Park Commissioners held a competition for the park at Back Bay, but Olmsted's firm did not enter. Olmsted agreed to review the accepted conceptual plan, and found it lacking for practical reasons – the plan did not adequately deal with the tidal flooding in the Back Bay. Subsequently, the Park Commissioners hired Olmsted to modify the plan and take over as the landscape architect for the Park Commissioners' project. Frederick Law Olmsted's work, augmented by the introduction of his nephew, John Charles Olmsted, into the firm as a partner in 1884, led to the planning of "The Parkway" in 1886. The Parkway connected Boston Common to Franklin Park via Commonwealth Avenue to Back Bay Fens along the Muddy River through a series of ponds to Jamaica Pond and on to the entrance to Franklin Park. The entire Parkway system required a comprehensive plan for taking back portions of marshes and reinforcing riverbanks. It also took plenty of mental fortitude from Olmsted to penetrate the political system that did not always allow comprehensive plans to be fully carried out.[50]

The most influential initiative proved to be the Muddy River Improvement. The park and parkway design that ultimately emerged grew out of a need to rectify a serious pollution problem as well as to shore up a series of riverbanks. By 1880, the Muddy River had changed from a reasonable stream to brackish water that was a spawning ground for mosquitoes and disease. Further, the filling in of the Back Bay Fens had put an end to the tidal flow that had previously cleared out the Muddy River, and sewage often piled up.[51] Once fixed according to Olmsted's plan, clean water and a stable riverbank led to increased economic investment and the promotion of good neighborhood development.[52]

In 1881, the Olmsted firm submitted John Charles Olmsted's "General Plan for the Sanitary Improvement of Muddy River," which built upon a previously submitted design by Frederick Law Olmsted, entitled, simply, "Suggestion." Boston and Brookline dragged their feet in implementing the plan for a number of reasons. The Muddy River served as the dividing line between Boston and Brookline, and action on the plan required the two jurisdictions to work in concert. It was not until 1892, when the Olmsted firm submitted another revision, that any progress took place. This "Plan of the Parkway Between Muddy River Gate House and Jamaica Park" clearly laid out the sanitary concerns that spurred on the project to begin with. It called for cleaning up the river and reinforcing the riverbanks, along with the construction of a parkway along the Muddy River to Jamaica Park.[53] Again, most notably, it was environmental and economic concerns that justified the Muddy River Project, rather than aesthetic considerations.

Charles Eliot (1859-1897), the designer of a series of parks and parkways in the Boston metropolitan area, built upon Frederick Law Olmsted's earlier work in Boston. Eliot brought the concepts of the park and the parkway outside the city and into the region, anticipating and attempting to plan for the growth of the metropolitan area. In a sense, Eliot's work in the Boston area was the foundation for the metropolitan vs. regional planning debate, discussed in later chapters.[54]

Charles Eliot, the son of Charles William Eliot, a Harvard professor and later its president, took a degree from Harvard in 1882. Immediately afterwards, he enrolled at Harvard's Bussey Institution and took classes in horticulture and agriculture. Before completing that course of study, he began an apprenticeship with Frederick Law Olmsted's firm in Brookline. The apprenticeship lasted from 1883 to 1885, whereupon Eliot completed his coursework at the Bussey Institution and embarked on a tour of European landscapes – following in the footsteps of Olmsted, Downing, and others in the new field of landscape architecture. He followed guidebooks and travelogues of European sites, with some additional help from family contacts.[55] Upon returning from Europe in 1886, Eliot set up his own landscape practice in Boston.

Charles Eliot's work in the Boston metropolitan area demonstrated as much political and organizational talent as it did forward-looking design innovations. Eliot built upon the earlier work of the Olmsted firm and brought it to the point at which it could, as well as possible given the political climate, anticipate and plan for the growing Boston area. He began with a vision of comprehensive planning brought about only by political organization and did not envision a systematic and cross-jurisdictional parks project without the authority to complete it comprehensively.

Eliot's efforts to set up the Trustees of Public Reservations in 1890 and the Metropolitan Parks Commission rank as his greatest accomplishments. The Trustees of Public Reservations were to act as a receivership for land throughout the Boston metropolitan area, and they were to hold it for the good of the public. As Eliot described it, the Trustees would be "composed of citizens of all the Boston towns, and empowered by the State to hold small and well-distributed parcels of land free of taxes, just as the Public Library holds books and the Art Museum pictures – for the use and enjoyment of the public."[56] This concept arose out of two primary issues and anticipated the need for regional planning. First, Eliot foresaw the continued development of the Boston area and wished to provide an outlet from the creeping urbanization. Second, Eliot knew the troubles associated with getting separate jurisdictions to act in concert to promote the public good derived from park development.[57] The Trustees of Public Reservations provided the foundation for cross-jurisdictional action. In a letter to Governor Russell of Massachusetts, Eliot called for "the establishment of some central and impartial body [...] empowered to create a system of public reservations for the benefit of the metropolitan district as a whole."[58]

Eliot worked with the Appalachian Mountain Club to create grassroots support for the legislation establishing the Trustees of Public Reservations. In May 1891, Russell signed the law establishing the Trustees and Eliot was named their secretary. Almost immediately, Trustees of Public Reservations picked up on Eliot's ideas and proposed four projects. First, the Trustees would survey and publish an inventory of public open spaces in Massachusetts. Second, they would compile and publish all the laws relating to public open spaces. Third, they called for a meeting with the leaders of the several park commissions in the region. And last, they called on the Massachusetts General Court (the legislature) to study the park circumstances.[59] According to Norman Newton, the final two proposals led to the establishment of "the first metropolitan system of parks in America," the Metropolitan Park Commission.[60]

Eliot's initial work with the Metropolitan Park Commission in 1892, as Landscape Architect, predated only slightly his re-association with the Olmsted firm, as a partner in March 1893. As the landscape architect for the Metropolitan Park Commission, Eliot oversaw the acquisition of parkland throughout the Boston area. By 1895, the Park Commission had acquired numerous holdings, including Middlesex Fells, Blue Hills, Revere Beach, and the Hemlock Gorge.[61]

Eliot consistently worked in a comprehensive manner to acquire land and develop it as parks. That is, the park system grew according to a plan that accommodated the elements of the natural landscape unveiled through surveys and careful study of the natural resources. This allowed for the public open space to weave together the city and the growing metropolitan area. Connecting the reservations were parkways that Eliot planned and designed.

The Metropolitan Park Commission's role further evolved out of the Boulevard Act of 1894, under which the Metropolitan Park Commission was to spend $500,000 buying land and constructing parkways to provide access to the open spaces. While initially resistant to the Commission's intervention and meddling in this area, Eliot later warmed to the idea of the parkways and incorporated them into the general plan, despite the fact that it required extra time to survey and reconfigure the plan.[62] His father, Charles W. Eliot, noted that at least part of the impetus behind the Boulevard Act was to provide jobs to the unemployed men of Boston, out of work due to the depression of the 1890s.[63] Finally, not only did the younger Eliot embrace the idea of parkways, but he also called for the incorporation of public transportation systems to shuttle the "masses" to and from the preserves.[64] Eliot's regional park system plan could not be criticized as elitist, as other regional park plans would be over the next fifty years or so. Further, access through public transportation added to urban reform efforts previously limited to the urban core.

Eliot determined that incorporating parkways into the "general plan" had to meet the same standards as any other element of the landscape design. That is, the parkway must exist as an integrated element. As a jobs program, the construction of parkways had to be continually in sync with the idea of the general plan. A survey of the land and its resources had to predate the planning and construction of a parkway. Eliot cited Frederick Law Olmsted when he proposed separating the parkway concept from the concept of the commercial highway in the landscape – one most likely constructed in the pursuit of commerce, not recreation or scenery.[65]

Eliot's work with the Trustees of Public Reservations and the Metropolitan Park Commission emerged as a seminal step in the development of parkway systems. First, Eliot moved to what Olmsted seemed to know at an earlier time: parks and parkways were best planned comprehensively and across local jurisdictional lines. Second, Eliot's plans required an understanding of the future growth of urban areas in the planning for suburban or, indeed, metropolitan park systems. Both of these concepts demonstrated concern for reform in the urban landscape. That is, as understanding of the growth of urban populations took shape, the urban reformers who still believed in the reformative benefits of parks and parkways aimed to adapt those same benefits to the region. This view conceptualized the region outward from the city and not from the region inward to the city.

2.3 Transformation from Urban to Suburban

David Schuyler, in *The New Urban Landscape*, addresses a transition (from city to region) in the urban park and parkway movement that occurred near the end of the nineteenth century, when urban park and parkway initiatives developed more comprehensively.[66] While the urban parks and parkways had evolved into part of the urban fabric by the end of the nineteenth century, the parkways continued to serve the city even as they grew outside the core city. After the experience with Boston and the Metropolitan Park Commission, the initial suburban-scale parkways brought about a new concept of urban reform – environmental reform as embodied in the Bronx River Parkway. Although more suburban in scale, this second set of parkways still was produced as a response to urban issues – from inside the city and outward.

2.3.1 Automobility

The increased usage and availability of the automobile played perhaps the most significant role in the transition of the park and parkway movement from the urban scale to the suburban scale. In 1900 there were 8,000 automobiles registered in the United States. By 1910, there were 458,300. Automobile use continued to grow at an exponential rate until by 1920, there were 8,131,500, and by 1930, there were 23,034,700 registered automobiles in the United States.[67] Although the automobile was not universally affordable, the rapid growth of the automobile between 1910 and 1920, in particular, did include growth in middle-class automobile ownership.[68]

2.3.2 Beginnings of the Suburban Parkways – Bronx River Parkway

The Bronx River Parkway (1895-1925 and beyond),[69] which was born of the city, began to spread the concept of reform outside the urban core and into the suburbs. The original construction spanned a 15-mile stretch from the Bronx Botanical Gardens to the Kensico Dam near Valhalla, New York.[70] It bridged the gap between the "parkway" of Olmsted and Vaux with its modification by Eliot in Boston, and the suburban commuter parkway that was so recognizable and integrated with the New York metropolitan landscape.

The Bronx River Parkway initially was a response to the environmental decline of the Bronx River. Not only had the adjacent houses and buildings contributed to pollution problems, but development along the river had led to severe erosion of the riverbank. In a project similar to Olmsted's Muddy River Project, where shoring up the riverbank allowed for a permanent road adjacent to the river while simultaneously preventing further riverside development, the Bronx River Parkway served a number of purposes.[71]

The Parkway provided automobile owners with a recreational outlet – the width of the right-of-way varied between 200 and 1200 feet for fifteen miles. The design provided for only limited access and therefore a constant flow of traffic.[72] Built in the Bronx River Valley, the limited-access Bronx River Parkway did not interfere with the daily transaction of commerce. Local roads merely passed over the parkway at convenient intervals.[73] The grade separation also established a sense of place for the Parkway. Limited entrances helped to locate the Parkway in a natural reserve. Finally, the Bronx River Parkway, using the automobile as catalyst, introduced the concept of the suburban parkway as a tool for economic development in the

New York region. Land adjacent to parkway land witnessed a many-fold increase in property value.[74]

The parkway itself served a number of functions aside from being the transportation corridor for which we now know most parkways. Further, the early parkway did not intrude on the landscape, but aimed to rehabilitate it while also providing for other uses – recreation, transportation, and suburban development facilitator.

The success of the Bronx River Parkway, spurred on in part by the growing number of automobiles on the road, brought about a strong constituency for additional parkways. By the early 1930s proponents of parkways around the country looked in near unison to Westchester County (north of Manhattan) as the model to follow when developing suburban parkways. Popular magazines published "before and after" photos and promoted the parkway's benefits.[75] Many later parkway projects were staffed with engineers, planners, and landscape architects who had experience on these early projects.[76]

These new suburban parkways served at least two distinct yet complementary functions. First, they existed as pleasure drives with integrated recreational facilities nearby. Second, they served as a thoroughfare for a growing suburban commuter population. The historiography associated with the urban parkway does not cleanly separate the urban and suburban parkways. However, the Bronx River Parkway, as stated above, served the City of New York as a reform-oriented project. Its success as a recreational outlet and as a facilitator of traffic allowed it to grow into the promising commuter conduit it ultimately became. But in its transition to commuter road, the Bronx River Parkway as an artifact of urban reform – in the vein of Olmsted, Vaux, and Eliot, among others – ceased to fulfill the promise of the early park and parkway movement. It no longer brought nature into the built environment of the city. The Bronx River Parkway effort attempted to allow the urban built environment to penetrate the yet-to-be-settled countryside surrounding the city by giving the city automobile owner access to the countryside.

In the first decades of the twentieth century, landscape architects and planners continued to plan and design roads for parks and natural areas, especially in the western United States, as well as parkways for growing metropolitan areas and decentralizing cities. In addition, increasing automobile use for the first time made vast expanses of the country accessible to the automobile tourist. Ultimately, it was an intersection of several elements that amassed a constituency for the regional parkway – tourists' desire for greater access to natural holdings, the compatibility of the parkway and the landscape, and the availability of the automobile. The very existence of the parkway or park road (that is to say, the issue of whether it should be built), its composition, its place in the landscape, and its design became matters of controversy, as discussed in the next chapters.

2.3.3 Discussion of Access to Public Lands

In 1916 Henry S. Graves, Chief of the U.S. Forest Service, wrote an article in the journal *The American City* entitled "Road Building in the National Forests." Among his primary concerns was accessibility for tourists:

> Each year sees an increasing use of the National Forests by residents of the more densely populated districts, east and west, who wish to escape the heat and discomfort of the city during the summer months [. . .]. It will be an increasingly important part of the work of the Forest Service to care for and render accessible these playgrounds of the nation.[77]

Graves believed that new roads would open up great tracts of land to the public. He also argued that roads would bring previously inaccessible areas into contact with the rest of the economy. That is, roads would open up a means of communication for isolated private landholders and facilitate the economic development of these lands separated from markets.[78]

Graves notes the inclusion of National Forest tracts and their access roads within the so-called "Park-to-Park Highway" project. The partially built Park-to-Park Highway provided access to the National Forests and National Parks in the western United States. Many sections of this highway, combined with road-building projects in the National Forests, provided access "to tourists and freighting."[79] So while considering the benefit to the automobile tourist, the Forest Service never took its eye off its primary function. Graves wrote:

> The construction of roads into such a territory would not only be of the greatest aid to the present settlers, but in many instances would make accessible new bodies of timber. Moreover, it would often open up to the general public important recreation possibilities.[80]

The National Forest road-building project had at least two sides: access to the forests for tourists and access to the economy for citizens living within National Forest boundaries.

2.4 Conceptualizing the Parkway before the Regional Debate

Frederick Law Olmsted Jr., John Charles Olmsted (nephew and adopted son of Frederick Law Olmsted), and Charles W. Eliot 2nd (nephew of Charles

Eliot), the offspring of the two most prominent park and parkway builders of the end of the 19th century, continued to work and write about parkways at the beginning of the automobile age. Each contributed to the discourse on parkways as the automobile emerged as an integral part of the landscape. John Charles Olmsted, with his cousin and stepbrother, continued the Olmsted tradition as a park designer. Similarly, Charles W. Eliot 2nd worked and wrote as a landscape architect in the early part of the 20th century.

In 1916, before the completion of the Bronx River Parkway, John Charles Olmsted outlined in *Landscape Architecture* the different types of parkways and their uses. Olmsted classified parkways in two ways: formal and informal. Formal parkways included boulevards such as Eastern Parkway in Brooklyn. Informal parkways such as the Bronx River Parkway, practically a park in and of itself, provided the user with a variety of natural scenery.[81] He wrote:

> Informal parkways (meaning more or less informal), curvilinear pleasure traffic routes, especially such as include or adjoin pleasing natural landscape features, should be much more generally adopted in suburban and rural districts than has been the practice, because, in proportion to cost, they are capable of affording much more pleasure than are formal boulevards, both to those who pass along them and those who live adjoining or near them. In fact, they frequently serve more or less completely as local parks.[82]

Moreover, the informal parkway allowed the road to lie more naturally upon the land. Designers could adjust to topography without the "excessive cuts and fills" required in the more formal parkway or boulevard.[83]

In a statement that reflected Frederick Law Olmsted Sr.'s reasoning and thought process forty or so years before, John Charles Olmsted noted the increase in property values associated with the more formal parkway. He wrote, "Formal boulevards are generally preferred to informal parkways by the real estate men because they require less land and because they combine easily with the usual rectangular subdivision with its gratifying implication of ultimate high city values of lots [. . .]."[84] A decade later, the Bronx River Parkway proved that property values also increased when developed adjacent to a more informal parkway.

In 1922, Charles W. Eliot 2nd wrote of the impact of the automobile on the parkway concept. In his article, "Influence of the Automobile on the Design of Park Roads," published in *Landscape Architecture Magazine*, Eliot further differentiated between types of park roads and parkways, while he also made suppositions regarding the future use of suburban and regional parkways. Changing technology – the passing of the horse-drawn carriage and the institutionalization of the auto – caused a "revaluation of the various factors in the design of park roads."[85] The natural park preserve, or "informal parks," as Eliot refers to them, stood to be most impacted by the automobile.[86]

The informal park had originally been designed, according to Eliot's understanding of Frederick Law Olmsted Sr., for "those unable to walk or ride horseback."[87] Existing park roads allowed visitors to experience the landscape and scenery in the slow-moving horse-drawn carriage. The speed of the automobile changed the scale on which a visitor could enjoy the landscape. Eliot noted that the automobile allowed "a great expanse of open country [to be] accessible to the automobile owner." Further, the "whole countryside has become the motorist's park."[88] Industrializing (through automobility) the countryside at a point just after his literal and philosophical brethren had built careers by bringing the natural environment back into the urban fabric seems quite contrary.

Eliot 2nd continued his proclamation of the countryside as the realm of the auto owner by contrasting it with the city. He wrote, "[i]t would seem fair under these conditions that the parks in the city should be designed primarily for the use of those to whom this open country is not accessible and that the pedestrian and horseman should have prior consideration."[89] In the region, Eliot 2nd viewed the automobile as privileged – if only because of its ubiquity. He wrote:

> While park roads in the days of carriages were incidental to the broad landscape effects of the park, and were subordinate elements in the design, the requirements of an automobile road are such that whatever may be done, it can not be made subordinate: it dominates its neighborhood.[90]

At this point, Eliot 2nd did not foresee the automobile's ability to similarly "dominate" the region. Urban parks, liberated from the automobile, satisfied Eliot 2nd and provided a foundation for the debate over the newly accessible region.

3. CONCLUSION

The view of the development of the urban park and parkway – from the city outward – while notably true when the history of the decentralized metropolitan United States is viewed, sidestepped or plain ignored the idea that the region could be viewed in the exactly opposite way. The legacy of the early park reform movement again was of the incorporation of the natural environment into the urban landscape. Industrialization had brought about the need for reform. After having been brought up in the cradle of this reform movement, Charles Eliot 2nd's call for automobility in the region seems, in retrospect, antithetical to earlier theories. The opening of the rural and the regional landscape to the automobile set the stage for the conflict out of which the debate over the regional parkway emerged. Is the region the place where the tools of city planning are used to construct an urban fabric, or is the region a place where the tools of regional planning are used to value small towns and rural economies?

---

[1] Norman T. Newton, *Design on the Land: The Development of Landscape Architecture* (Cambridge, MA and London, England: The Belknap P of Harvard U, 1971) 596.

[2] Christian Zapatka, "The American Parkways," *Lotus International* 5 (1987): 96-128. See also, Charles William Eliot, *Charles Eliot: Landscape Architect* (1902; Freeport, NY: Books for Libraries P, 1971) and John Nolen and Henry V. Hubbard, *Harvard City Planning Studies, Volume XI: Parkways and Land Values* (London: Oxford U P, 1937).

[3] Zapatka, "The American Parkways" 97.

[4] See for instance, Richard E. Foglesong, *Planning the Capitalist City: The Colonial Era to the 1920s* (Princeton: Princeton U P, 1986).

[5] See Zapatka, "The American Parkways," and Ethan Carr, *Wilderness by Design: Landscape Architecture and the National Park Service* (Lincoln and London: U of Nebraska P, 1998) 97-127.

[6] See, for example, David Schuyler, *The New Urban Landscape: The Redefinition of City Form in Nineteenth-Century America* (Baltimore and London: Johns Hopkins U P, 1986), Foglesong, *Planning the Capitalist City*, and Sam Bass Warner, *The Urban Wilderness: A History of the American City* (New York: Harper & Row, Publishers, 1972) among others.

[7] See Schuyler 3-4.

[8] See Thomas Bender, "The 'Rural' Cemetery Movement: Urban Travail and the Appeal of Nature," *New England Quarterly* 67 (June 1974): 197.

[9] Bender 196.

[10] See Schuyler 37.

[11] See John W. Reps, *The Making of Urban America: A History of City Planning in the United States* (Princeton: Princeton U P, 1965) 326. Reps wrote: "The popularity of these rural cemeteries for uses other than as burial places must have astounded and perhaps horrified their sponsors."

[12] Reps 326.

[13] Downing's writings from the *Horticulturist* were reprinted posthumously in a single volume. A. J. Downing, *Rural Essays*, ed. George William Curtis (1853; New York: Da Capo, 1974) 144.

[14] Downing 141.

[15] Schuyler 51-53.

[16] See a broad set of histories, including Newton 267ff; Schuyler 75ff; Roy Rosenzweig and Elizabeth Blackmar, *The Park and the People: A History of Central Park* (Ithaca: Cornell U P, 1992) 15ff; and Francesco Dal Co, "From Parks to the Region: Progressive Ideology and the Reform of the American City," *The American City: From the Civil War to the New Deal*, Giorgio Ciucci, Francesco Dal Co, Mario Manieri-Elia, and Manfredo Tafuri, translated from the Italian by Barbara Luigia La Penta (Cambridge: MIT P, 1979) 143-292.

[17] Rosenzweig and Blackmar 18.

---

[18] See Newton 267ff. Also, see Rosenzweig and Blackmar 121ff, Dal Co 160ff, Carr, *Wilderness by Design* 20ff. There is an extensive set of Olmsted biographies including: Frederick Law Olmsted Jr. and Theodora Kimball, eds., *Frederick Law Olmsted: Landscape Architect* (New York and London: The Knickerbocker P, 1928) and Laura Wood Roper, *FLO: A Biography of Frederick Law Olmsted* (Baltimore and London: Johns Hopkins U P, 1973).

[19] See Rosenzweig and Blackmar 130-132, Carr, *Wilderness by Design* 11-20, and Dal Co 160-164.

[20] Dal Co 163-164. Dal Co mentions only Olmsted in this passage seemingly omitting Vaux. The omission of Vaux occurred often with historians as well as with Olmsted in his own writing.

[21] Christian Zapatka, *The American Landscape*, edited by Mirko Zardino (New York: Princeton Architectural P, 1995) 31.

[22] See Rosenzweig and Blackmar 132.

[23] See George F. Chadwick, *The Park and the Town: Public Landscape in the 19th and 20th Centuries* (New York and Washington: Frederick A. Praeger, 1966) 184-85, and also Schuyler, for the concept of the separated natural and urban environments.

[24] See Chadwick 185.

[25] See Ethan Carr, *Wilderness by Design* 17-18.

[26] See Zapatka, *The American Landscape* 31-32.

[27] Schuyler 195.

[28] Dal Co 164.

[29] See Zapatka, *The American Landscape* 31, and also Newton 275-278.

[30] Olmsted, Vaux, & Co., "Report to the Brooklyn Park Commission," Brooklyn, January 8, 1868, reprinted in *The Papers of Frederick Law Olmsted: Writings on Public Parks, Parkways, and Park Systems*, Supplementary Series Volume I, ed. by Charles E. Beveridge and Carolyn F. Hoffman (Baltimore and London: Johns Hopkins U P, 1997) 112-146. Reprinted in the Papers of FLO as "The Concept of the 'Park-Way.'"

[31] Newton 276.

[32] Zapatka, *The American Landscape* 31 and Schuyler 123.

[33] Zapatka, *The American Landscape* 31 and Schuyler 121-123.

[34] Olmsted, Vaux, & Co. 114.

[35] Olmsted, Vaux, & Co. 125.

[36] Olmsted, Vaux, & Co. 127.

[37] Olmsted, Vaux, & Co. 129.

[38] Olmsted, Vaux, & Co. 135-137.

[39] Zapatka, "The American Parkways" 99-101, and Schuyler 128. Also, Ethan Carr, "The Parkway in New York City," *Parkways: Past, Present, and Future* (Boone, NC: Appalachian Consortium P, 1987) 121-128.

[40] Schuyler 128.

[41] Carr, "The Parkway in New York City" 122.

[42] Zapatka, "The American Parkway" 103 and Schuyler 133.

[43] Schuyler 129ff.

[44] Schuyler 129.

[45] Charles Beveridge, "Buffalo's Park and Parkway System," *Buffalo Architecture: A Guide*, ed. Reyner Banham (Cambridge, MA: MIT P, 1982) 15-23.

[46] Beveridge, "Buffalo's Park" 15-19.

[47] Beveridge, "Buffalo's Park" 15-19.

---

[48] Beveridge, "Buffalo's Park" 21.
[49] Schuyler 138-47, Zapatka, "The American Parkway" 103-105; Newton 290-95, 318-23.
[50] Schuyler 138-47, Zapatka, "The American Parkway" 103-105; Newton 290-95, 318-23.
[51] Cynthia Zaitzevsky, *Frederick Law Olmsted and the Boston Park System* (Cambridge, MA: Belknap P of Harvard U P, 1982) 82.
[52] Zaitzevsky 84.
[53] Zaitzevsky 83-84.
[54] Zapatka, "The American Parkway" 105. An analysis of the landscape – especially its scale – indicates this idea.
[55] This description comes primarily from Newton 290-306, Schuyler 138-144, and Frederick Law Olmsted, "Public Parks and the Enlargement of Towns" reprinted in *The Papers of Frederick Law Olmsted: Writings on Public Parks, Parkways, and Park Systems*. It is meant to set the stage for the Boston work of Olmsted's young partner, Charles Eliot.
[55] Eliot 132. See also Newton 318-336.
[56] Eliot 318.
[57] See Newton 318-336.
[58] Eliot 356-357.
[59] See Newton 323.
[60] Newton 323.
[61] See Newton 331 and Eliot 738-741.
[62] Newton 331. Newton notes that Charles Eliot felt parkways were beyond the scope of the original Metropolitan Park Commission mission.
[63] Eliot 456.
[64] Eliot 463.
[65] Eliot 656.
[66] Schuyler 146.
[67] See *Historical Statistics of the United States, From Colonial Times to 1970*, Part 2 (U.S. Department of Commerce, Washington, DC: GPO, 1975) 716.
[68] See Paul S. Sutter, "Driven Wild: The Intellectual and Cultural Origins of Wilderness Advocacy During the Interwar Years (Aldo Leopold, Robert Sterling Yard, Benton MacKaye, Bob Marshall)" diss., U of Kansas, 1997, 41.
[69] See Marilyn E. Weigold, "Pioneering in Parks and Parkways: Westchester County, New York, 1895-1945," *Essays in Public Works History* 9 (February 1980): 1-43.
[70] See Zapatka, "The American Parkway" 110-113.
[71] See Carr, "The Parkway in New York City" 122, and Weigold 1.
[72] Zapatka, "The American Parkway" 110-113.
[73] Newton 600.
[74] Weigold 11.
[75] Newton 600.
[76] See Carr, "The Parkway in New York City" 123.
[77] Henry S. Graves, "Road Building in the National Forests," *The American City* 16:1 (1916): 4.
[78] Graves 5.
[79] Graves 5.
[80] Graves 6.
[81] John Charles Olmsted, "Classes of Parkways," *Landscape Architecture*, vol. 6 (Oct. 1915 – Jan. 1916): 37-42.
[82] John Charles Olmsted 42.

---

[83] John Charles Olmsted 43.
[84] John Charles Olmsted 38.
[85] Charles W. Eliot, 2nd, "The Influence of the Automobile on the Design of Park Roads," *Landscape Architecture Magazine* 13 (October 1922): 27.
[86] Eliot, 2nd 27.
[87] Eliot, 2nd 28.
[88] Eliot, 2nd 28.
[89] Eliot, 2nd 28.
[90] Eliot, 2nd 29.

Chapter 3

Regional Visionaries

Benton MacKaye, Lewis Mumford, the Regional Planning Association of America, and the Region

> We need the big sweep of hills or sea as tonic for our jaded nerves – And so Mr. Benton MacKaye offers us a new theme in regional planning. It is not a plan for more efficient labor, but a plan of escape. He would as far as is practicable conserve the whole stretch of The Appalachian Mountains for recreation. Recreation in the biggest sense – the re-creation of the spirit that is being crushed by the machinery of the modern industrial city – the spirit of fellowship and cooperation.[1]
>
> *Clarence Stein, 1921.*

1. INTRODUCTION

The history of the regional parkway in the eastern United States is intimately tied to the development of the Appalachian Trail. This chapter discusses the theory of regional planning brought out by the development of the Appalachian Trail concept, the broader debate over regional planning, and the legacy of this theory, which in the end brought on the conflict associated with the development of the regional parkway.

The main players in this story are the forester/regional planner Benton MacKaye (1879-1974) and the American cultural critic and regionalist Lewis Mumford (1895-1990). It was MacKaye's Appalachian Trail article, published in the *Journal of the American Institute of Architects* in 1921, that began the debate between the regional planners associated with the Regional Planning Association of America (RPAA) and the "metropolitan" planners associated with Thomas Adams (1871-1940) and the Regional Plan of New

York and Its Environs (RPNY & E). This debate culminated in the 1932 publication of one of the essential polemics in the planning field – Lewis Mumford's criticism of the Regional Plan of New York and its Environs – and Thomas Adams' defense of that plan. The Appalachian Trail project provided a practical alternative to the developing urban- and suburban-based automobile culture and that culture's offspring – auto tourism and road development. The broad debate – a vision in the face of decentralization versus accommodation to the market – defined the two sides in the conflict over the development of the parkway in the region.

Benton MacKaye's article "An Appalachian Trail: A Project in Regional Planning" was part of the theoretical and practical model for a vision of regional planning in the United States. The vision included an orderly recentralization throughout the region, that is, regional planning that respected small-town culture and values, valued work and leisure in the same place, and attempted to keep from the region the life-draining afflictions of congestion and sprawling "dinosaur cities." This vision, according to MacKaye, would help to reverse the migration from rural to urban, mitigate the desire to merely escape the overcrowded cities, facilitate sustainable agricultural and forest industries, and provide opportunities for recreation, rejuvenation, and healthful living.[2] MacKaye wrote, " [i]t is the slow quiet development of a special type of community – the recreation camp. It is something neither urban nor rural. It escapes the hecticness of the one, the loneliness of the other."[3] Over time, the boosters of the national park in the Southern Appalachians, the Skyline Drive, and the Green Mountain Parkway would use some of MacKaye's rhetorical vision in support of their own projects. As will become apparent, however, these boosters really only understood part of MacKaye's plan.

Through the RPAA, which began in 1923, a loosely organized group of planners, architects, housing experts, economists, and intellectuals fleshed out a regional planning vision sparked by MacKaye's Appalachian Trail article. Lewis Mumford served as the primary spokesperson for the group and articulated the regionalist view best, certainly better than MacKaye did.[4]

The regional vision advocated by MacKaye and Mumford called for regional economic development on a scale compatible with the inhabitants of the region. That is, it called for development in line with the existing and potential agricultural and forest resources. The regional development approach devised by MacKaye presented an alternative to unbounded industrialization evident in the cities and metropolitan areas. Re-entry, recentralization, or even migration into the region at the appropriate scale was a humanizing force, one that would, in the future, allow for a reconnection of living and working in a partnership that had been severed by industrialization. The primacy of the connection between human beings and the land, as outlined by MacKaye, led to the regional visionaries' rejection of the parkway as idealized and constructed by the metropolitan planners.

2. ROOTS OF THE REGIONAL VISION

2.1 Patrick Geddes and the Influence of the Regional Survey

Benton MacKaye and Lewis Mumford subscribed to many of the ideas espoused by the Scottish biologist/planner Patrick Geddes (1854-1932). Geddes' ideas influenced the conception of the region, regional planning, and the regional vision supported by MacKaye and Mumford. Peter Hall, in *Cities of Tomorrow*, characterized Geddes as "an unclassifiable polymath who officially taught biology (more probably, anything but biology)."[5] Moreover, not only did Mumford champion many of Geddes' ideas, he articulated and wrote about them in a way Geddes could not.

According to Frank G. Novak Jr., the editor of *Lewis Mumford and Patrick Geddes: The Correspondence*, Geddes influenced Mumford in three major areas. First, Geddes' view of the city, its historical make-up and future potential, as related through his 1915 book *Cities in Evolution*. Geddes' analysis helped Mumford understand the relationship of the city to the region. Second, Geddes taught Mumford to apply his intellectual findings as an activist. Finally, Geddes served as Mumford's "model of the intellectual as generalist."[6]

Both Lewis Mumford and Thomas Adams invoked Patrick Geddes in support of ideologically opposite approaches to regionalism. This fact is testimony to the breadth of Geddes' scope and to the murkiness of his discourse, since Geddes' writings were not always clear. Geddes' two primary influences were his scientifically based survey and his political/economic understanding of history. The historical analysis quickly came out as the major influence on Mumford, MacKaye, and others. The survey methodology emerged as the second among equal influences. On the other hand, it was Thomas Adams and the RPNY & E who invoked the Geddesian survey, without the historical underpinnings, to justify their work.

Geddes viewed the industrial age and the innovations in new technology as a perfect opportunity to push society towards a more equitable and efficient use of land and wealth. Just as the Paleolithic and Neolithic eras of the Stone Age delineated two phases of human progress, the Industrial Age seemed to demonstrate that it was prepared to make the next step to the Neotechnic order, to use Geddes' terms, out of the Paleotechnic era of mixed priorities and inefficiencies. The construction of the Paleo/Neo terms demonstrated an avowed optimism for the progress of a society overwhelmed with constant innovations. According to Geddes, the advances were to be used in positive and progressive ways. Further, by virtue of the survey, the path to reform existed and society only needed a nudge in the correct direction.[7]

Geddes used the term "conurbation" to define the sprawl of urbanization associated with the growth of great cities during the Paleotechnic era (mid nineteenth century industrialization). In this era, the primary interest was the maximization of profits and increased production, without concern for efficiencies. Conurbation caused the inefficient use of land, resources, and the available technology. Paths of growth followed the markets, which in turn followed the profits. The valley survey demonstrated that, historically, growth had followed flows of people, of water, and of natural resources. By virtue of innovations in technology, civilization was now poised on the edge of a new era where industrialization could be used to promote the progress of society, not its deterioration. In the Neotechnic era, the public was charged with maximizing efficiencies in the use of land and resources and reversing the private dissipation that had previously occurred in the Paleotechnic era.[8]

According to Geddes and his work, *Cities in Evolution*, the "survey" was the basis for planning – regional or otherwise. This concept grew out of Geddes' knowledge of nineteenth-century French geography.[9] The survey's scope included all the natural, cultural, and man-made resources existing in the region. The survey, which Geddes also referred to as the "Valley Section," described the way of life, the way things worked in the region.

To counteract the possibilities for future conurbations, Geddes proposed a return to the valley survey, which revealed that the industrial age produced modes of flows – roads, railroads, etc. – which should be used to delineate paths of development rather than merely facilitate amorphous sprawl. "Town extensions naturally extend star-wise along main thoroughfares. They can be kept from growing together by placing schools, playgrounds [...] in unbuilt areas left between."[10] To Geddes, the survey, historical analysis, and understanding of the available technological innovations made the issue of regionalism clear. For the practitioners and theorists who understood Geddes as interpreted by Mumford, the rhetoric emerged as similarly clear.

Geddes possessed an ideological bias towards the centralized control of planning that had no connection to political awareness. Rather, it came from his scientific and social survey. He analyzed history, picked up on earlier works such as George Perkins Marsh's *Man and Nature*, and pushed for logical, efficient uses of land and resources. It is this basic, simple approach that appealed to both Mumford and Adams, who each employed pieces of Geddes' work.

Geddes surveyed the region from the city outward. Specifically, he surveyed the region from the Outlook Tower in Edinburgh. Mumford described it best:

> The Outlook Tower is both a real building and an idea. It stands on Castlehill, at the head of High Street in Edinburgh, watching over that historic mile between the Castle and Holyrood, where the events of Scotch history are sealed in a hundred buried stones and living edifices [. . .]. From the gallery at the top of the Tower one has a view of Edinburgh and the surrounding region [. . .].[11]

The survey's importance was clear and the survey had innate usefulness for regional planning. Mumford's earliest writings on regionalism recognize the importance of the city as a cultural artifact. And although he wrote in the May 1925 edition of *The Survey* that, "the hope of the city lies outside itself," the city remained central to the region. The hope of the future of the city rested with orderly and efficient decentralization that would relieve congestion. Later, Times Square replaced the Outlook Tower as the point of reference for Mumford, yet the view was still the same.

The survey rooted in geography, science, anthropology, history, and an understanding of the natural resources – revealed the plan. In 1915, when Geddes published *Cities in Evolution*, the City Beautiful Movement still reigned in the United States. To Mumford and others concerned with city planning, the survey concept did not fit within the generally understood concept of the City Beautiful.[12] While the concept of the "survey" seems obvious today, during the period following the World's Columbian Exhibition in 1893, Daniel Burnham did not utter the words, "survey, then make no little plans," but proclaimed loudly, "make no little plans."[13]

In addition to the survey concept, Geddes' method of analysis allowed for Mumford to understand the roots of a truly American culture – the roots which Mumford would attempt to uncover in *The Golden Day*. Notably, Mumford found those roots in a regional context. And finally, the analysis of the growth of cities into the region, the flow of people out of settlements, and the recognition of the impact of technologies on regional change all influenced Mumford's conception of what he termed "migrations."[14] Mumford's initial article in the Regional Planning Number of *The Survey Graphic* began with an homage to Geddes. Mumford wrote:

> The great migrations that swept over Europe in the past; the migrations that surged past the water-boundaries of Europe and crawled through the

> formidable American wilderness – these great tides of population, which unloosed all the old bonds, have presented such an opportunity.[15]

Mumford, as he was to continue to do for MacKaye, clarified sometimes difficult to understand rhetoric.

2.2 The Garden City Movement and the Region

In addition to Patrick Geddes' conception of the region, MacKaye and Mumford looked to the English Garden City ideals of Ebenezer Howard. The Garden City concept combined with the Geddesian notions of the region to formulate the basic tenets of the regional vision. Ebenezer Howard first published his concept of the Garden City in 1898, under the title *To-morrow: A Peaceful Path to Real Reform.*[16]

Howard's concept of the Garden City evolved out of his understanding of a number of reformist ideas studied and pursued in Victorian England. As Peter Hall notes in *Cities of Tomorrow*, the Victorian city was considered a horrific place for all but society's most well-off. Similarly, the English countryside was not much better due to the depressed agricultural economy and the flight of population to the cities. The Garden City idea combined numerous concepts put forward by others in response to congestion, the recentralization of industry in already overcrowded cities, and the economic inaccessibility of adequate housing for England's ever-increasing working class.[17]

The Garden City emerged as a unique concept because of the way Howard picked up on the most plausible of the concepts put forward by late 19th century reformers. The Garden City incorporated the ideas of a planned community, municipal ownership of land and housing (profits from rising rents in times of economic expansion were to be used for community betterment), neighborhood or ward units arranged around a civic focal point, definitively bounded "urban" areas, a population threshold, an industrial belt, and a greenbelt to define the boundary and provide agricultural space for self-sufficient food production. Howard's work deliberately left out other reformist concepts, including the idea that change could occur within the present form of the city (a metropolitan reform) and that "municipal control" should also include controlling the means of production and distribution.[18]

Despite Howard's concerted effort to escape the "utopian" label cast upon other reformers, even the most benevolent reviewers of his self-financed publication of *To-morrow* could not get past the idea that new cities, planned or otherwise, were just not practical. The concepts of new towns still were associated with the more radical reformers such as the

Englishman J. Bruce Wallace and the American Albert Kinsey Owens. Other reformers like Henry George and Edward Bellamy had argued that reform could and should occur within the reality of the ever-expanding existing city. Regardless of the criticism of Howard's vision, two existing new towns in England, Bourneville and Port Sunlight, were examples of at least part of the idea in practice. Moreover, Buder, Hall and others point out that Howard (while in the United States during the mid-1870s) must have seen Frederick Law Olmsted's garden suburb of Riverside, near Chicago.[19]

Soon after the Garden City Association (GCA) began construction of the first garden city, Letchworth, GCA Secretary Thomas Adams and Raymond Unwin began to move town planning away from Howard's Garden City ideal and towards garden suburbs (as in Hampstead garden suburb, ca 1907), suburbia in general, and town extension. Howard's fundamental Garden Cities ideas did experience a revival through the work of F.J. Osborne and C.B. Purdom before and after World War II, only to be derailed by Howard himself when he proposed Welwyn in England in 1919 (which was more of a garden suburb).[20]

2.3 The Garden City in the United States

In part, the concept of regionalism in the United States grew out of the Garden City theory of Howard and the belief that the regulatory restrictions that emerged before WWI were not a strong enough response to the deteriorating conditions in American cities. Housing reformers, the playground movement, labor reformers, and other social reformers banded together with business interests intent on preserving property values in cities to form and enact the earliest zoning regulations, based on the German model. Progressive Era reformers had championed these use restrictions, along with an understanding of reform through the Garden City idea. However, the ability to enact zoning ordinances and tenement reform regulations, and the inability of the short-lived Garden City Association of America (founded 1906) to accomplish anything, shifted the Progressive Era reformers' interest away from Garden Cities. Moreover, as Richard Foglesong has noted, the earliest American use of even the most basic of the English Garden City concepts, at Forest Hills Gardens in 1913, only indicated that Garden City theories could work well and protect the middle class from speculative profit-making and rent gouging. The protection of the middle class from landowners and developers was not the main concern of the social reformers of the pre-World War I era.[21]

The ideas on reform circulating just after WWI presented housing activists, planners, land reformers, and environmentalists with a multitude of ideas rather than a unifying one. To the housing reformers, the tenement restrictions were inadequate because of their minimum standards. They called on planners to bring about proactive prescriptions for the development of housing and solutions to the overcrowded situations in urban areas. Further, many had begun to anticipate the impact that the auto would soon have on connecting the urban and rural environments. WWI had opened up the debate over publicly planned and funded housing, with the possibility of removing the ability of speculators to obtain windfall profits at the expense of those most in need of reasonably priced, adequate housing. Finally, the changes in industrial processes, as understood and dealt with during the war effort, had exposed the wasteful use of natural and human resources. Again, a unifying theme was needed to project solutions to ongoing changes that seemed to have multiple causes and possibilities. The concept of regionalism stepped into this void as the needed unifying theme.

Regarding the fractured possibilities, the myriad tools available, and their inadequacy, Lewis Mumford wrote in 1919, "The single tax by itself has only fulminated against the tenement. Housing reform by itself has only standardized the tenement. City planning by itself has only extended the tenement."[22] Regionalism, influenced by the Garden City ideal, developed as the unifying theme that tied the issues together.

2.4 The Golden Day

The final source of influence on Mumford's regional conceptualization is his perspective on history as portrayed in *The Golden Day*. The era prior to the Civil War and the rise of industrialization had existed as the cultural and historical high point in the United States, according to Mumford. Mumford championed the time of Emerson, Thoreau, and others with specific reference to "communal form most evident in New England."[23] The cultural strength of the United States existed in its past. The form of this strength was literally the small New England town and village. Mumford could visualize a return to this era through the incorporation of the regional vision, which evolved as a picture through the methodology outlined by Geddes – the survey and the structured historical view. The future of the vision emerged, literally, in the form of Garden Cities.

Mumford published *The Golden Day* in 1926. Prior to that book, Mumford had brought out a number of the ideas on regionalism and the cultural high point of the United States evident in *The Golden Day*. In his 1924 article "Devastated Regions" in *The American Mercury*, Mumford presented the issue of regionalism clearly, paid homage to those who influenced him, and gave a glimpse of what was to come in the subsequently published Regional Planning Number of *The Survey Graphic*. In "Devastated Regions," the "friend" responds to the "critic":

> It works both ways: we have made the city and the country equally intolerable: neither is suited to permanent human habitation [. . .]. A hundred years ago there was plenty of life in my region [. . .]. Today a few big dairy farms are all that keep the folks going; the inhabitants who have remained manage to scrape along in the crudest fashion.[24]

In Mumford's view, the region is a place of permanence. It is a place of sustained economic development, of human-scale towns and cities (the Garden City in the region), and it is a place of culture – farm culture, timber culture, and human culture.

3. THE REGIONAL VISIONARIES

3.1 Lewis Mumford

> In a period of flows, men have the opportunity to remold themselves and their institutions.[25]

So wrote Lewis Mumford in the May 1925 issue of *The Survey Graphic*, devoted to regional planning. The publication of the "Regional Planning Number" of *The Survey Graphic* was the primary intellectual outlet for the members of the Regional Planning Association of America. Mumford based his call to action on the several migrations of the population in the United States. The term "migration" described the historical patterns of settlement within the context of natural-resource use. The first migration was the clearing of land west to the Allegheny Mountains. It resulted in a wasteful use of forests and land. The second migration reordered settlement upon the land. This reordering was made possible by the railroad and by industrialization along rivers (taking advantage of the opportunities for water as a source of power) and left the land and human resources ruined. The third migration came with further advances in railroad transport and the innovation of coal power, populations re-centered around industrial urban areas and financial centers. The third migration caused overcrowding in the cities and drained resources from the outlying factory towns yet aggregate cultural resources increased, as did opportunities for connections overseas. On the brink of the fourth migration, Mumford correctly predicted another dispersal of population, brought on by increased innovations in technologies.[26]

The fourth migration would open up two avenues, according to Mumford. One would merely follow the examples of the first three migrations towards inefficient use of land, technologies, and resources. The other avenue called for recentralization in small towns and regional cities, along with decentralization of congested industrial urban centers.

Mumford called for a type of regional planning that would lead, rather than merely follow. Decentralization defined the fourth migration, yet to Mumford and the other regionalists, a well-planned decentralization would exist as an end in itself. That is, the fourth migration, if left to market forces, would only require a further adjustment in land settlement. With regional planning as a guide, the fourth migration would lead to a stable, more sustainable settlement pattern.

MacKaye followed a similar argument in *The New Exploration.* He championed the efficient use of resources, whereas he considered inefficient uses illogical and costly. As with Mumford, standing on the threshold of a new era (MacKaye used the term "folk flow") provided MacKaye with a regional solution supported by scientific survey, economic and political realities, and a historical context. Planning for regionalism was not changing the way people did things, but rather finding the most efficient path through a survey of the resources in the region.[27] The analogy used by MacKaye was that finding a way over a mountain is a matter of finding the switchbacks that already exist, rather than making a new path directly over the top. Further, since populations were beginning to flow out of cities into the region, the regional planner needed to find the natural path, rather than the most direct and destructive.[28]

For Mumford, who viewed the region from the city outwards (as opposed to MacKaye who took his view from the mountaintop back to the city), the regional vision grew out of three primary sources. They were Patrick Geddes, the Garden City movement, and, as an intellectual and historical concept, the cultural roots of America – small towns, human scale economies, and a relationship to the landscape.[29] Mumford viewed the region as the past and the future of cultural America. He opened his second article in *The Survey Graphic*'s Regional Planning Number with an optimistic call for regionalism – "the hope of the city lies outside of itself."[30] This sentence paid homage to Geddes, the regional planning roots in Garden Cities, and the cultural realities that had brought American civilization to the 1920s.

3.2 Emile Benton MacKaye

In his introduction to MacKaye's 1928 discourse on regional planning, *The New Exploration*, Lewis Mumford compared Benton MacKaye's work to that of Henry David Thoreau and George Perkins Marsh.[31] This description is quite remarkable, given the fact that many of MacKaye's ideas were more understandable when Mumford articulated them, yet the fact that

Mumford took the time to clarify MacKaye's views is a testament to the importance of the ideas, if not their expression.

Benton MacKaye was born in Stamford, Connecticut in 1879, the fifth of six children and the youngest son. He lived his early years in Manhattan, where his father, James Morrison Steele MacKaye (known as Steele), worked as an actor, playwright, producer, and stage designer. At the age of nine, MacKaye moved with his family to the place that would be one of the primary influences on his life and work, "an old, almost untouched New England village, Shirley Center [Massachusetts] [. . .]."[32]

MacKaye's experiences in Shirley Center influenced nearly all of his lifelong ideas. The New England village had all the qualities of community, place, and environment necessary for living the good life. The decline of the colonial New England villages, "'drained' to augment the 'waters' [the growing metropolitanism],"[33] was the driving force behind MacKaye's concept of regional planning.

The village of Shirley Center served as the model community in MacKaye's view, in which there existed three "elemental environments" – the primeval, the rural and the urban.[34] The model community's historical demise caused by the ever-expanding metropolitan area framed the problem that MacKaye spent his life attempting to solve. Shirley Center's existence in the rural environment and proximity to the "primeval" (MacKaye's conceptualization of the wilderness as "the environment of life's sources, of the common living-ground of all mankind"[35]) was the foundation for his lifelong pursuit of "geotechnics." MacKaye defined geotechnics as "the applied science of making the earth more habitable." Shirley Center's success and continued well-being depended on keeping the rural places rural, the urban places urban, and the primeval wild. Habitability depended on the quality of each place in the region.

In his early life, MacKaye spent many hours and days in the woods near Shirley Center. This activity prepared him for his later work as a forester and as a strong proponent of the survey – the task of learning and recording the natural systems evident in the environment. One of MacKaye's explorations became known as "expedition nining" and is best described by Mumford in the introduction to *The New Exploration*:

> In his characteristic systematic way, MacKaye laid out a whole series of walks and dignified them, again with thoughts of Humboldt, by calling them "expeditions." On one of these expeditions, he not merely accompanied his United States topographic map with a brief narrative, but tore loose with philosophic reflections in a language all his own. This happened to be his ninth expedition; and from that time on his brother James, who discovered Benton's opus and liked to recite high-

> flown passages from it with great glee, used to call this broader kind of nature study "expedition nining."[36]

Mumford continued by describing MacKaye's life as one long "expedition nine."[37]

The expedition and surveying activity, along with residence in Shirley Center, influenced MacKaye in many ways throughout his lifetime, which he spent championing the outdoor life, both personally and to others. Also influential were Harvard, where MacKaye studied geography and forestry, and the work of Patrick Geddes.

In the 1950 article "Growth of a New Science," published in *The Survey*, MacKaye cites the work of the Harvard geographer William Morris Davis and Patrick Geddes as the "top guideposts of my own working career."[38] In this article MacKaye, in a romantic way, recalled from memory Davis' opening statement in the initial "Geography A" lecture: "'Gentlemen,' said he [Davis] [. . .], 'here is the subject of our study [holding up a six-inch globe] – this planet, its lands, waters, atmosphere, and life; the abode of plant, animal, and man – the earth as a habitable globe.'"[39] Davis' method of teaching geography through the exploration of the earth reflected MacKaye's own experience of exploring the area in and around Shirley Center. MacKaye found a way to apply his "expedition nine" to a greater scale.

MacKaye graduated from Harvard in 1900 and returned in 1903 to pursue the newly offered discipline of forestry, until 1905. Many years later he wrote, "Forestry may be defined as the practice of growing woods instead of mining them – one of the first great applications of science to make substantial portions of the earth 'more habitable.'"[40] His training led him to the U. S. Forest Service, where he worked under Gifford Pinchot. It also introduced him to public service, in which he worked in various agencies, off and on, until 1920. Prior to World War I, MacKaye lived in Washington, DC, and worked in an environment described as "a clearinghouse for new reform proposals and a collecting-point for bright young bureaucrats and publicists intent on implementing them."[41] Among those MacKaye met during this time period was the economist Stuart Chase, who would become his colleague in the RPAA during the 1920s.

In 1912, MacKaye worked for the U. S. Geological Survey (USGS), mapping and measuring forests in the White Mountains of New Hampshire. Work with the USGS broadened MacKaye's understanding of the relationship between land, water, and planning.[42] The USGS work caused MacKaye to view the world more broadly, and it forced him to work with other experts in a bureaucracy.

By 1919 MacKaye had left the Forest Service and transferred to the Department of Labor headed by Secretary Louis F. Post (a single-tax proponent and friend of Henry George), which published MacKaye's report, *Employment and Natural Resources*. The study's publication grew out of an immediate concern for men and women "dislocated" during the war years – soldiers returning from abroad and domestic workers uprooted for the war effort. MacKaye, however, had been working on the broader issue of resettlement for a number of years.[43] In this work, MacKaye applied the idea of habitability and related social concerns to concepts previously devoid of such concerns. In 1950, commenting on this report, MacKaye explained that true employment opportunity in forestry required permanence more closely associated with agriculture than "forest mining." He also cited the difficulty of preventing land from becoming the object of speculation. MacKaye wrote of the need to "enable the settler to make a farm out of stump land without becoming a speculator himself or the victim of one."[44]

During these years in and out of public service, MacKaye brought to bear a social side of ecology, environment, and conservation. While the USGS survey of the White Mountains mapped and charted the natural systems associated with the flow of run-off through forested lands, *Employment and Natural Resources* factored in the human systems associated with natural-resource management. It was at this point that MacKaye had moved beyond conservation and natural-resource management in and of itself and into the realm of regional planning. Regional planning, for MacKaye, was a direct engagement with economic issues and thus was within the mainstream of planning ideology. Moreover, he approached regional planning from the perspective of the region looking back towards the city, considering all regional resources, rather than approaching regional planning from the viewpoint of the city looking outward.

Forestry school had taught MacKaye the "sustainable" side of growing trees. Forestry was agriculture in this new discipline, not the extraction of a non-renewable resource. Unfortunately for foresters, the schools of forestry turned out professionals who understood the science of growing cycles and diseases, but did not understand the plight of the workers engaged in forestry. In an article entitled "Some Social Aspects of Forest Management," MacKaye introduced the problem in this way:

> Recent symptoms of unrest among the timber workers in the Northwest and elsewhere have revealed a new problem for the American forester. It is the problem of the lumberjack. Our forest schools in their processes of turning out foresters have courses in silviculture, mensuration, dendrology, protection, influences, management, utilization, lumbering, etc; but the lumberjack himself and the very human problems that go

> with him do not occur in the curriculum. It may require a later age to reduce these matters to the textbook, but out in what we call "real life" they must be faced without waiting for book knowledge.[45]

"Forest-culture" required the professional forester to understand the social aspects of employment, community building, and the industry's connection to the rest of the region.

MacKaye's discussion of forestry within the context of the "very human problems" that go along with the industry revolve around community, stability, transportation, and a "dependable and long-time form of ownership."[46] Further, since the plight of the lumberjack mirrored that of the farmer and agricultural worker, MacKaye's analysis included workers using land and natural resources as their source of employment. The 1919 study, *Employment and Natural Resources*, applied the issues facing the lumberjack to all the workers in natural resources. At this point in MacKaye's career, he thought the responsibility for creating a stable community, and giving the agricultural worker a fair chance at stability, rested in part with the government, through the reservation of land for communal purposes and in part with coordinated community settlement.[47]

A number of the major concepts that became the foundation for MacKaye's "Appalachian Trail" article existed in his own mind (or at least the seeds for these concepts existed in his mind) by the time the Department of Labor published *Employment and Natural Resources*. The primary concepts were five-fold. First, the model town, Shirley Center, exhibited all of the traits for habitability – community, human scale, and proximity to the wilderness. On the downside, Shirley Center could not resist urbanization (MacKaye used the phrase "metropolitan flood" and characterized it as a rising "tide"[48]) that overwhelmed its resources and employment opportunities (this concept became more apparent as automobile use grew during the 1920s). Metropolitan expansion was the antithesis to the small town, and metropolitan growth contributed to the demise of small towns.

The second concept was the experience of the land as demonstrated through "expedition nining" and the ideas associated with surveying, as MacKaye had done as a child in Shirley Center and as a public servant working for the Forest Service and the Geological Survey. Third, MacKaye's training as a geographer and forester reinforced his ideas on conservation, on the one side, and the need for understanding the "social aspects" of forestry and resource management on the other. Fourth, MacKaye's further study of the employment issues associated with the broadest conception of our natural resources, primarily land, led him to champion cooperation in settlement patterns, land-use, transportation, and management of production. And, finally, and perhaps most importantly,

MacKaye saw these issues, geographically, as problems of the region and not of the city. The solutions to the issues raised by these primary concepts existed in regional planning.

By the end of 1919, MacKaye had lost his job with the Labor Department, and later his short-lived stint with the Postal Service was not successful.[49] By 1921, MacKaye had moved to Manhattan and worked as a writer for a newspaper. That spring he lost his wife, Jessie Stubbs MacKaye, who after a long mental illness apparently committed suicide. Under these conditions, MacKaye spent some time with Charles Harris Whitaker, editor of the *Journal of the American Institute of Architects*, writing and thinking. At Whitaker's farmstead in Mt. Olive, New Jersey, MacKaye conceived of and articulated his "Appalachian Trail" idea.[50]

3.2.1 An Appalachian Trail – MacKaye's 1921 Article

The *Journal of the American Institute of Architects* published Benton MacKaye's "An Appalachian Trail: A Project in Regional Planning" in October 1921 (Figure 3.1). As the title indicates, MacKaye's "Appalachian Trail" is two things – a trail with recreation at its core (MacKaye's term of "re-creation" is best) and a study in regional planning. In his introduction, Clarence Stein made this evident: "To all those to whom community or regional planning means more than the opening up of new roads for the acquisition of wealth, this project of Mr. MacKaye's must appeal. It is a plan for the conservation not of things – machines and land – but of men and their love of freedom and fellowship."[51]

Stein's words regarding regional planning, and the fact that it encompassed more than merely the facilitation of the acquisition of wealth, described the crux of MacKaye's argument. MacKaye championed not only regional planning *vis-a-vis* the beneficial effects of leisure time and the scouting life, but also regional planning as a way to promote sustainable economic development and recreational activities. Just as *Employment and Natural Resources* promoted "timber farming," and not "timber mining," "An Appalachian Trail: A Project in Regional Planning" promoted a plan for sustainable living, and not tourism or other resource-draining activities.

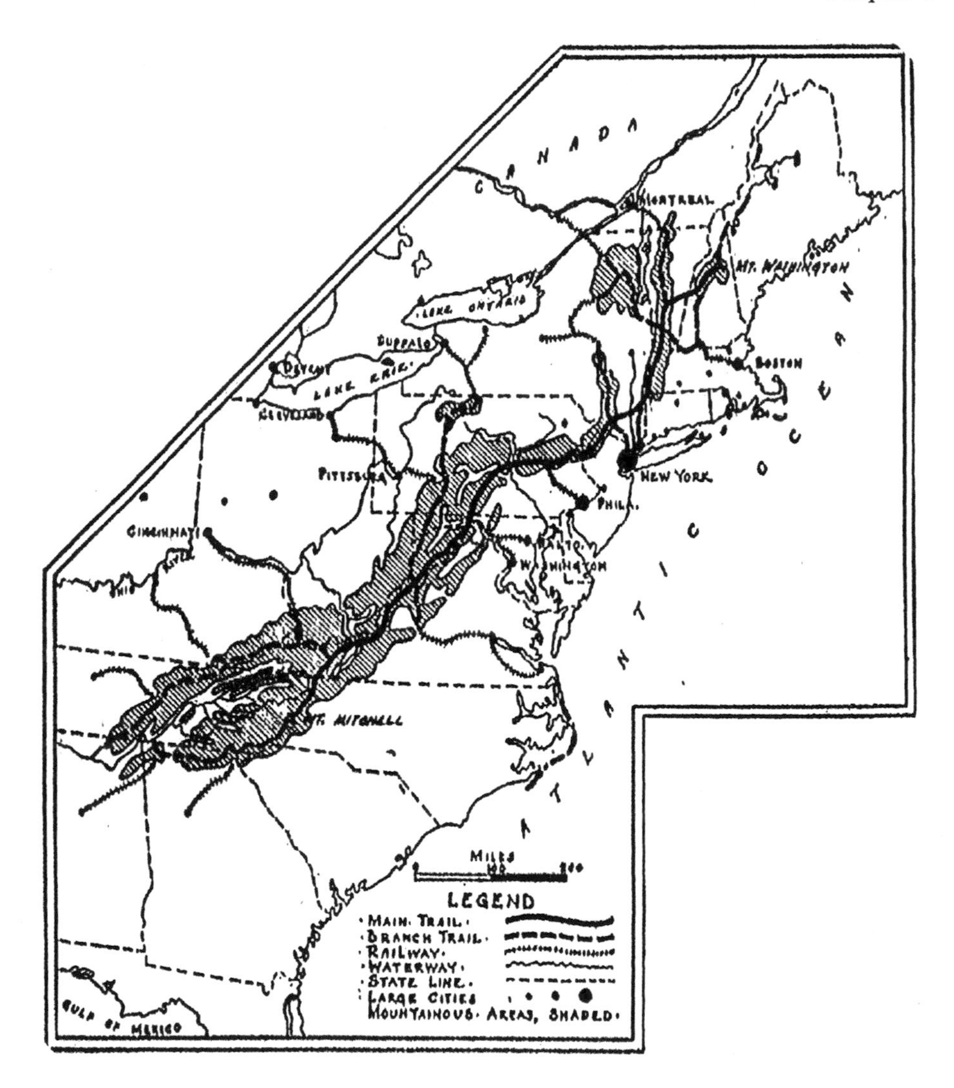

Figure 3.1. An Appalachian Trail. Benton MacKaye, "An Appalachian Trail: A Project in Regional Planning," *Journal of the American Institute of Architects* IXX (October 1921): 7.

The *JAIA* article explained many of the themes MacKaye held aloft as important aspects of life: regional planning, repopulating the Appalachian region, regional development (as a proactive initiative and as dam against the flood of urbanization), natural systems and resources, employment, the positives associated with recreation and the "scouting" movement, the Appalachian Range's importance as a geological reference point on the East Coast, and the need to prevent the draining of the regional population and wealth by the metropolitan areas.[52]

MacKaye's treatment of these issues served as the foundation for regional planning as a concept. Similarly, these concepts explain MacKaye's later writings on regional parkways, both in support of and in opposition to them. In sum, "An Appalachian Trail: A Project in Regional Planning" tied together many of MacKaye's views that had been essential to his professional and personal life well prior to the publication of the article. It further was the starting point for his later work in regional planning and wilderness conservation.

MacKaye began the Appalachian Trail article by revealing the growth of the recreation-camp movement in the United States. The scouting movement complemented the recreation-camp and together they began to solve the "problem of living" associated with the complexities of modern society. He warned, however, that society need not return to an era before industrialization or to a time before the existence of the tools that made life easier in order to benefit from the experience of the natural world.[53] The value in the outdoor life is more than merely escaping the hecticness of progress and the economic ups and downs associated with modern society. The therapeutic value in the scouting life needed to have an associative meaning brought about by the cooperative camps. To MacKaye, experiencing the land alone was not as valuable as acting on the land and developing it in a way that promoted cooperation among people, and cooperation between the land and people. This was the experience of the pre-industrial, colonial town in New England.

3.2.2 Regional Planning and Employment

As a blueprint for regional planning, MacKaye's article described the opportunities extant in the Appalachian Range through the eyes of "a giant standing high on the skyline along these mountain ridges, his head just scraping the floating clouds."[54] In 1921, MacKaye had yet to meet Patrick Geddes, so the survey technique may not be traceable to Geddes at this point; however, the concept of visualizing the entire region is fundamental to MacKaye's proposal. MacKaye's giant finds three opportunities in the north to south survey of the Appalachians. First was the opportunity for recreation for the people of the overpopulated East Coast:

> [The giant] recollects vast areas of secluded forests, pastoral lands, and water courses, which, with proper facilities and protection, could be made to serve as the breath of a real life for the toilers in the bee-hive cities along the Atlantic seaboard and elsewhere.[55]

The second opportunity MacKaye simply described as recuperation. Most likely still severely affected by the death of his wife, MacKaye noted the recuperative powers of the "mountain air along the Appalachian skyline."[56]

The third opportunity, one left out in many scholarly discussions of MacKaye's Appalachian Trail work, was for employment in the Appalachian region. The discussion in the *JAIA* article mirrored the *Employment and Natural Resources* discussion of the flow of population from the rural to the urban and the impact of that on the economy of the rural areas. The rural population of the United States declined from 60% in 1900 to 49% in 1920.[57] In the Blue Ridge Mountains of Virginia, the decline in population during this period was even more dramatic. The pull of employment opportunities, the chestnut blight, and the decreasing fertility of farmland all contributed to the out-migration.[58] MacKaye's call for regional planning as a solution to the economic problems associated with the rural areas of the Appalachians is analogous to his call for better employment opportunities in the "stump country" of the upper Midwest. At this point, MacKaye did not call for preservation of pristine nature. He advocated thoughtful use of the land without exploitation.

It is arguable that the economic reality of the 1920s Appalachian residents reflected the experience of many other populations also tied to the land. The modern economic reality had left these people in Appalachia with strong cultural roots, yet they could not be valued because of economic circumstances, which made them vulnerable to the rising tide of urbanization. MacKaye's regional planning ideas sought preservation of indigenous culture (Shirley Center, Appalachia, etc.) by virtue of new economic terms conducive to its survival.

The Appalachian Trail project attempted to set out these new terms through regional planning. MacKaye did not advocate removing people from the land in order to create playgrounds for visitors from the cities. He advocated repopulating of the rural environment so that people could make use of the land, live within the context of the land, and take part in the three opportunities envisioned by the giant on the skyline: recreation, recuperation, and employment. The repopulating policy needed to make use of some of the strategies outlined in *Employment and Natural Resources* and other early writings. In the *JAIA* article, MacKaye referred to these strategies as the "new deal" in agriculture. MacKaye noted the existence of a number of National Forests along the Appalachian Mountain chain. The public ownership of these lands in the East was central to MacKaye's vision of recreation, recuperation, and employment.[59]

To reverse the flow of population from rural to urban, MacKaye proposed cooperative recreational communities along the Trail. Urban workers would initially devote idle or leisure time to the projects. MacKaye believed in the value of permanent community camps along the Trail to "stimulate every possible line of outdoor non-industrial endeavor."[60] MacKaye proposed a Garden City-type solution to the problem of community size: "Greater numbers should be accommodated by more communities, not larger ones."[61]

The regional planning outlook of the Appalachian Trail project was meant to be organic and related to the experiences of those who took part in its creation. MacKaye believed the typical worker would commit two weeks per year to the project in an effort to escape the industrial grind. Two weeks per year "on the mountain top would show up many things about life during the other fifty weeks below."[62] Two weeks in the mountains engaged in recreation and recuperation would show off the difference between the "recreative and non-industrial life" on the one hand and the "industrial life" on the other.[63] While this distinction provided knowledge for workers who returned to the industrial centers at the end of two weeks, MacKaye believed the knowledge would also stimulate new conceptions for employment away from the cities. He wrote:

> The organization of the cooperative camping life would tend to draw people out of the cities. Coming as visitors they would be loath to return. They would become desirous of settling down in the country – to *work* in the open as well as *play*. The various camps would require food. Why not raise food, as well as consume it, on the cooperative plan? Food and farm camps should come about as a natural sequence. Timber also is required. Permanent small-scale operations could be encouraged in the various Appalachian National Forests. The government now claims this as a part of its forest policy. The camping life would stimulate forestry as well as a better agriculture. Employment in both would tend to become enlarged.[64]

The regional planning vision clearly built upon the ideas in *Employment and Natural Resources* as well as his short article, "Some Social Aspects of Forest Management."

3.2.3 "Social Reform and Social Readjustment": The Appalachian Trail's Precursor

In a 1921 memorandum entitled, "Regional Planning and Social Readjustment," MacKaye laid the foundation for his later *JAIA* article on the Appalachian Trail. In this memo, MacKaye attempted to make a political and economic argument for a place-based social-reform movement. MacKaye's arguments were wide-ranging, anti-corporate, anti-war, and rhetorically difficult, at best. However, this memorandum provided the structure for the Appalachian Trail concept published in October 1921. [65]

MacKaye proposed an employee-based industrial complex whereby workers would have the incentive to produce "for use and not for profit."[66] MacKaye termed this "buying over" by workers the "people's industry." Regional planning, MacKaye argued, would provide the place-based organization for this new industrial complex to thrive. This new form allowed for production and consumption, for employment and leisure, again, in the same place, without the pressures of the congestion-producing, life-draining realities of the urban industrialized model.[67] Although MacKaye called this new regional community a "compound community" (one model for the industrial community, one for the non-industrial community), he had really only recreated the idyllic colonial New England town. Moreover, despite suggesting a Garden City model for the industrial-compound communities, his rhetoric implied that he based his concept on the New England small town – Shirley Center. He conceived the non-industrial compound communities as human-scale camps of limited size. And just as the Garden City innovators had suggested, MacKaye called for new camps or new industrial communities, rather than larger camps or communities, to accommodate population growth.[68]

MacKaye believed the Appalachian Mountain range provided the best place for the manifestation of his regional planning and social-reform project. The Appalachians provided the largest piece of unpopulated or under-populated land in the East. Further, national and state forests made up a good portion of the land in the Appalachians and public ownership of large chunks of land was necessary for MacKaye's vision. He suggested three plans for carrying out his vision. First, he offered a plan for the regional development of industrial communities in Appalachia. He based the proposed camps on the availability of nearby resources, such as textile-manufacturing communities spaced throughout the Piedmont area of the Southern Appalachians to take advantage of cotton production. Second, MacKaye called for the development of non-industrial camps throughout the Appalachians. These camps, working in conjunction with the industrial camps, would provide recreational outlets to workers within urban areas. Like his subsequently proposed shelter camps, MacKaye proposed connecting the non-industrial camps with a footpath. Finally, the third plan consisted of constructing the Trail itself. According to MacKaye, each plan was interconnected and reliant upon the others, although he did call for the construction of the Trail first, so as to get workers into the region to experience the benefits of that place.[69] This memorandum certainly provided the foundation for the October 1921 Appalachian Trail article. Equally important, however, is the fact that the memo provided insight into MacKaye's thinking and further explained the interconnectedness between the Appalachian Trail and its social-reform meaning.

MacKaye confirmed that his project was committed to regional planning and the development of resources. He laid out the basic structure of the regional development plan and true social reform. It began with the Trail, built over time in sections. Those working on the trail and hiking would have access to shelter camps, again, built over time and as needed. Community camps were slated for construction after the shelter camps. And last, MacKaye envisioned volunteers constructing the food and farm camps for those committed to regional development of farm and forest communities. The food and farm camp development "would provide tangible opportunity for working out by actual experiment a fundamental matter in the problem of living."[70]

In the memorandum MacKaye explained the importance of the interconnectedness of each part of his "project in regional planning." He recognized that the social reform efforts of the community camps and the shelter camps required a new way of looking at production, industrialization, leisure, and recreation. He called it "social readjustment" because of the necessary and complementary reeducation process that had to occur. MacKaye calculated that the construction of the Trail prior to the establishment of the camps and shelters would lead the way towards "social readjustment." His thinking on this topic seemed to evolve as he prepared the memorandum, in which he presented the idea of the industrial and non-industrial camps first, and argued that they were each interconnected and their mutual existence depended upon each other. Presented last in the memorandum was the Trail idea:

> The locating and building of this trail would constitute the first piece of work to be undertaken. This should be accomplished, as far as possible, through the organization of volunteer workers.[71]

MacKaye then, in a bold assumption, relied upon the desire of these "volunteer workers" to begin the drive towards social readjustment. That is, he believed in the strength and benefit of the land in the region to draw volunteers out of cities to a newly "adjusted" social atmosphere that would serve as a catalyst for institutional social reform.

Without reference to the uniqueness of the Appalachian Mountain culture, MacKaye made an economic argument for reclaiming for the "folks" the valleys and mountainsides. The previous economic circumstances – timber mining, farming without access to markets, and the inefficient use of land – had pushed the Appalachian "folks" to the margin of the economic realities. MacKaye wrote of "Appalachian empire building" along the Appalachian Trail. Again, mirroring his prior work, he wrote in the February 18, 1923, edition of the *New York Times*:

> The opposite of forest mining is forest culture (or forestry). This cuts from the forest valley each year only as much timber as grows there each year. So the valley never becomes empty of timber, industry or folks [. . .]. Many an Appalachian hilltown has fewer families today than a hundred years ago. It has been wrecked through forest mining: it can be restored through forest culture.[72]

The economic reform associated with "sustainable" timbering throughout Appalachia, for instance, would lead to more stable regional communities.

In *Employment and Natural Resources*, MacKaye argued that temporary employment is just that: temporary and no solution to employment problems unless it leads to permanent employment. He wrote of employment opportunities on rural land devastated by out-migration and previously unsustainable extractive methods:

> Immediate employment of a temporary nature, on road construction or similar activity, would amount to a real opportunity, provided it be made to lead directly to some form of permanent occupation, such as a home upon the land.[73]

MacKaye, in proposing the Appalachian Trail project, never intended for tourism to exist as the primary employment outlet for the "folk" of the mountain communities. Further, he seemed disdainful of the preservation of historical artifacts without economic viability. This is quite evident in an exchange between MacKaye and Lewis Mumford in 1927. Mumford was critical of historical revivalism that led to reenactments, people dressed in colonial costume, and the tourism trade associated with the artificial preservation of rural historic past. Mumford wrote, "wearing the Colonial costume in masquerade has something of the paradoxical effect of Henry Ford preserving the Wayside Inn [. . .]. What we look for, as an alternative to metropolitanism, is not a revival of the old; it is a fresh growth of something new [. . .]." MacKaye, responding a few days later, agreed. He wrote, "[. . .] the indigenous is that which is *permanent* rather than that which is past [. . .]."[74]

3.2.4 Natural Systems

MacKaye's early writing, prior to the Appalachian Trail article, consistently reflected the idea that properly managed rural and regional development could coexist with primeval wilderness. Criticism of "timber mining" and other industries dependent on natural resources focused on methods, not their singular existence. *Employment and Natural Resources* and "Some Social Aspects of Forest Management" promote sustainable use of the forested and agricultural lands and mountains. The idea of sustainable use of the natural resources and its compatibility with conservation is a strong theme evident in MacKaye's early work. Further, it is important to remember that MacKaye was a trained forester and foresters were not generally advocates of wilderness preservation.

Notably, MacKaye did not write of the preservation of the natural system, wilderness, or forestlands at this point in his career. His forester roots influenced his belief in the development of natural resources, and he argued the case for natural resources to be used in an efficient and sustainable manner. MacKaye believed it necessary to find a pattern and method of development of natural resources that could provide both employment opportunities and leisure opportunities. This conceptualization had roots in Jeffersonian agrarianism and an ideology of decentralization, but it required some type of centralized cooperative planning. It was a vision that did not require the setting aside of land specifically for wilderness preservation, yet it conserved land.

By 1928, with the publication of *The New Exploration*, MacKaye changed his tone and began to write of wilderness, using the terms primeval and indigenous. The growing consumerism embodied by roads that nurtured metropolitan growth and the seeming demise of the regional vision caused MacKaye to write of damming the metropolitan flood. In 1928, he wrote:

> This levee follows the main mountain way – through the Hudson Highlands, along the Blue Ridge of Pennsylvania and Virginia and throughout the vastness of the great Carolina Highland. Here is the backbone of Appalachian America. Here is the barrier of barriers within the world-empire of industrial and metropolitan upheaval.[75]

Thus, by the late 1920s, stemming overwhelming metropolitan growth had equal standing for MacKaye with the proactive reform efforts of the regional vision. MacKaye began to value the wilderness for its ability to serve as a buffer to metropolitanism and worked to mitigate efforts to damage the integrity of this buffer.

4. THE REGION AND THE METROPOLITAN TIDE

By the late 1920s and the publication of *The New Exploration*, Benton MacKaye had turned from primarily championing regional development of natural resources to creating a barrier against metropolitanism. In a chapter of *The New Exploration* entitled "Controlling the Metropolitan Invasion," MacKaye equated growing metropolitan development with rising floodwaters. Just as water flows down valleys and along streams, the metropolitan invasion flowed out from the urban areas along highways, "distributing population in a series of strings, which together would make a metropolitan cobweb of the locality. In this way the area with its several villages would become engulfed by the metropolitan flood."[76]

Just as rising floodwaters required physical barriers, the metropolitan invasion required physical containment. MacKaye called for the employment of prominent topographical features as the barrier – the "wilderness areas."[77] The wilderness areas formed boundaries. The highways, or motorways, spread metropolitanism. Not only did the rising metropolitan tide need wilderness barriers to keep it in check, it required appropriately developed roads to keep the creeping metropolitanism within its designated channels.

In the 1925 article "Planning the Fourth Migration," Lewis Mumford argued that American society stood on the verge of a new regional settlement pattern. That is, the economic and cultural circumstances that brought people to the great urban centers of the industrial age had changed. The new circumstances, spurred on by changing technologies such as the automobile, and a new culture of decentralization, allowed for two avenues of development. The first avenue, unchecked metropolitan growth, would put the problems of the urban areas into the region. The second would, through regional planning, both return America to the cultural highlight – the "Golden Day" – and push America forward to a more progressive era, an era where technology would pull people up, rather than push them down.[78]

4.1 Roads and the Appalachian Trail

Roads were at the center of MacKaye's vision. In the region, roads served to get the products of natural-resource cultivation to their markets. They also served as the key component of the infrastructure necessary for resource development. Appropriately planned, roads were the foundation of sustainable employment in the region. Roads as facilitators of production differed from roads as recreation. Not until the publication of "Tennessee – Seed of a National Plan" in 1933 and "Flankline vs. Skyline" in 1934 did MacKaye write about the automobile and roads as forming together a recreational activity in and of themselves. Prior to these articles, roads in MacKaye's conception served to facilitate the development of regional resources.[79]

In *Employment and Natural Resources*, MacKaye explained the primacy of land in employment issues. As an initial step in the development of natural resources, especially agricultural and forestry, MacKaye called for the construction of roads to take products from the region to the city. This infrastructure would need to predate the construction of the community, the farmhouse, or lumberjack house. The road construction, in turn, would provide employment opportunities and facilitate resettlement.[80] MacKaye viewed the Appalachian Trail project as a way "to open up a realm,"[81] and opening the realm required transportation. Again, the planned roads were utilitarian – intended to serve the industries associated with the regional development.

In the early 1920s the recreational parkway in the region was not among the ideas inherent in MacKaye's formulation of regional planning, but the fact that MacKaye saw roads as useful, indeed essential, for the developing of the region suggests that the concept of the parkway developed naturally in his philosophy.

4.2 The Townless Highway or the Highwayless Town

Benton MacKaye wrote specifically about parkways and highways as limited-access transportation corridors for linking cities and towns. Together, Mumford and MacKaye put forward further analysis, centered on the concept of the experience of the natural environment.[82] Further, their concern over stemming unchecked metropolitan growth and methods for creating the barrier had much to do with their conception of the townless highway and recreational parkway.

For MacKaye and Mumford, the motorway/highway presented numerous problems, mainly the development of the "motor slum" and congestion. In the August 1931 edition of *Harper's Magazine*, Mumford and MacKaye wrote of this problem: "[. . .] consider the motor road as a long-distance form of transportation [. . .]. All the time that is saved in the country stretches is lost once the car enters the city streets [. . .]."[83] Both proposed bypass-type roadways called "townless highways" and highways with limited-development access to counteract these problems.

The "Townless Highway" concept was a way to curb the consumerism associated with the growing urbanization. The bypass road not only prevented the development of the motor-slum, it prevented the degradation of the existing community. The introduction of the highway into towns, villages, and cities did as much damage as the development along newly constructed highways. Mumford and MacKaye termed the idea of excluding the highway from the town the "Highwayless Town." Of course, the "Townless Highway" connected the "Highwayless Towns." Limited-access highway development and regional planning that included new towns modeled on Garden Cities and Radburn, New Jersey, remedied the consumerism characterized by "frontage developments – the peanut stand, the hot-dog kennel [. . .]."[84]

MacKaye and Mumford presented three principles as part of their "Proposal for the Automobile Age." First, in order to avoid the development of the motor-slum, roads connecting places would need restricted access, just as the railroads had done with railway stations. This concept grew out of historical analysis and MacKaye's previous work on "Townless Highways."[85] But, according to Mumford and MacKaye, the through road needed more than merely a bypass route to mitigate roadside development and provide the necessary buffer from neighborhoods. The authors cited the need for open space and the positive examples set by Massachusetts in the development of the Mohawk Trail, as well as the work of the Bronx Parkway Commission and the Westchester County Park Commission. The Bronx River Parkway and the early Westchester County Parkways limited roadside development opportunities[86] and simultaneously provided open space for the driver and a buffer for the nearby neighborhoods.

Finally, MacKaye and Mumford argued that all through roads had to be parkways. That is, these roads needed to exist on their own, segregated from local traffic, on a preserve that provided open space to the driver and a buffer to the adjacent neighborhoods, and the roadway needed to follow the topography of the land.[87]

Mumford and MacKaye recognized the impact of the automobile to an extent greater than many others wrestling with the problem at the time. The "Townless Highway" concept demonstrated that Mumford and MacKaye believed the automobile could lead to needed decentralization without the pitfalls of metropolitanization. In the *Harper's* article, they cited Radburn and the Westchester Parkways as positive examples of decentralization.[88] They viewed the limited-access nature of the Westchester system as the answer and believed that the park preserves and open space would prevent sprawling development between access points along the parkway. Similarly, they believed that parkways, which could connect Radburn-like developments, would lead to similar decentralization without metropolitan congestion. While to this day both Radburn and the Westchester Parkways remain positive examples of the garden suburb and the suburban parkway, respectively, metropolitan sprawl did in fact consume each. Moreover, neither Mumford nor MacKaye would have found any solace in the idea that Radburn and the Westchester Parkways are, in 2002, better than the surrounding examples.

4.3 Flankline versus Skyline and Skyline Drives

The recreational parkway – skyline-type drives in particular – created a problem for MacKaye. The parkway, by his own admission, was a "strip of land devoted to recreation [. . .]."[89] Yet automobile recreation, in and of itself, ran contrary to his understanding of recreation at this point in his career. Over time, MacKaye grudgingly conceded the need for roads to accommodate the auto-recreation constituency.

MacKaye's writing was always clear on the method of developing the primeval. His Appalachian Trail study privileged the foot traveler, and the Appalachian Trail project built on at least three things. First, the idea of regional hiking trails in the Appalachians had existed since the early part of the twentieth century.[90] Second, the Trail idea had origins in his own experience growing up in New England, where, according to Mumford, MacKaye recounted his walks through the countryside by assigning physical and philosophical descriptions to each route.[91] And, finally, MacKaye believed the Appalachian Trail experience would lead to a better understanding of the role of the environment and natural resources in regional development.[92]

The proliferation of the skyline-type drive, to MacKaye, was a misuse of the automobile at the expense of the hiker. MacKaye continually supported the idea that the privileged user of the skyline had to be the pedestrian/hiker, even as supporters of the Skyline Drive and the proposed Green Mountain Parkway looked to parkways as facilitators of MacKaye's attempt to get people out of the city and into the primeval. His understanding of the parkway/trail controversy is laid out in an article entitled "Flankline vs. Skyline," published in a 1934 issue of *Appalachia*.[93] He reiterated his argument in a note written from Knoxville in July 1934. Regarding the Appalachian mountains, MacKaye wrote:

> This critical American wilderness is at present jeopardized as it never was before. I refer immediately to the slashing open of our eastern wilderness belts by skyline roads, specifically by the proposed parkways for the White and Green Mountain Ranges in New England and the Park-to-Park Highway in the Southern Appalachians [. . .].[94]

The flankline versus skyline debate emerged as a rhetorical compromise for MacKaye. As discussed in the chapter on the Green Mountain Parkway, the designer of the Green Mountain Parkway attempted to follow MacKaye's guidelines emphasized in the *Appalachia* article. In the end, however, MacKaye's flankline concession was only a temporary divergence from his trajectory towards wilderness advocacy. His privileging of the wilderness ultimately left little room for the automobile.[95]

5. CONCLUSION

This chapter provides a foundation for the regional vision. As the two most prominent voices for the regionalist view, Mumford and MacKaye sought to put forward an ideology that promoted decentralization into the region. This was combined with a social-reform platform that would have, if realized, significantly changed the dominant economic reality. The roots of the regionalist vision were firmly embedded in the fertile soil of late-nineteenth century English social reformers and Garden City advocates, as well as the town planning theory and philosophical underpinnings of Patrick Geddes. With these models, the regionalists recognized the impending decentralization brought about through changing economic times and new transportation technology, and advocated planning to guide an orderly recentralization in the region. The regionalists sought to value the small town, rural environment inherent in the Appalachian range. They aimed to facilitate rural economies, sustainable resource use, and long-term employment. Benton MacKaye's 1921 article, "An Appalachian Trail: A Project in Regional Planning," served as the foundation for the regional view in the United States. MacKaye's earlier memorandum, "Regional Planning and Social Readjustment," demonstrated the depth of the concept of social reform within the regional planning vision.

The regionalist view attempted to counter the rising metropolitan tide. The recognition of the path towards orderly, efficient regional planning assumed an impending decentralization with at least two options. New transportation technology tended to facilitate the spread of urban areas. This creeping metropolitanism threatened to change fundamental American values imbedded within the historical psyche of the American people. MacKaye and Mumford each viewed the nineteenth-century New England town as the locale that represented the quintessential place in American history. Through social reform and regional planning, they each proposed directions that led back to that place. While they did understand that positive values came from the past, neither Mumford nor MacKaye was a reactionary nostalgic. Noting MacKaye's and Mumford's deep understanding of Patrick Geddes buttresses the argument that each was future-oriented. Advances in technology, especially transportation technology, improved the prospects for the small town in rural America. Not only did each believe regional planning would stem the unchecked metropolitan growth, but they believed it would breathe life back into the rural areas slowly choked by the pull of urban congestion.

The primeval wilderness, still readily accessible to the nineteenth century small town, provided the recreational alternative to the congested city. The decentralization of city populations out into the region provided even better access to this important segment of MacKaye's conception of the twentieth-century reality. Implicit in MacKaye's understanding of the primeval was the idea of first-hand and intimate knowledge of it. Experiencing the wilderness required time, access, and the ability to escape from industrialization, modernism, and perhaps, even civilization itself. To MacKaye, this escape was, in and of itself, civilizing. The introduction of modernization, transportation technology, and the automobile into the primeval went against the very foundation of MacKaye's understanding of wilderness specifically, and his conception of the world, more broadly. Access to the primeval made cities, and even small towns, livable.

From this foundation of the regionalist vision, one side of the conflict that helped shape the regional parkway emerged. The regionalists wanted to lead a social reform movement. The metropolitanists, as discussed in the next chapter, wanted to accommodate the market and merely tweak it when it grew out of control. Certainly, both MacKaye's and Mumford's views evolved throughout the 1920s and 1930s. They perhaps became more pragmatic in their beliefs in regional planning. By the mid-1930s, MacKaye had made concessions to the recreational bases for the automobile and parkways. By then, however, he also realized the depth of his commitment to the wilderness ideal and worked to pursue that. This realization came about, in part, because of the attempts to put the parkway into the region.

---

[1] Clarence Stein, introduction, "An Appalachian Trail: A Project in Regional Planning," by Benton MacKaye, *Journal of the American Institute of Architects* 9 (Oct. 1921): 325-330.

[2] Benton MacKaye, "An Appalachian Trail: A Project in Regional Planning," *Journal of the American Institute of Architects* 9 (Oct. 1921): 325-330.

[3] MacKaye, "An Appalachian Trail."

[4] See John L. Thomas, "Lewis Mumford, Benton MacKaye, and the Regional Vision," *Lewis Mumford: Public Intellectual* (New York: Oxford U P, 1990) 67.

[5] Peter Hall, *Cities of Tomorrow* (Oxford, UK and Cambridge, MA: Blackwell, 1988) 138. Geddes "gave India's rulers idiosyncratic advice on how to run their cities and tried to encapsulate the meaning of life on folded scraps of paper."

[6] Frank G. Novak Jr., ed., *Lewis Mumford and Patrick Geddes: The Correspondence* (London: Routledge, 1995), 24. See also, Patrick Geddes, *Cities in Evolution: An Introduction to the Town Planning Movement and to the Study of Civics* (1915; New York: Harper Torchbooks, 1968).

[7] Geddes, *Cities in Evolution* chapter 2.

[8] See Geddes, *Cities in Evolution* chapter 2.

[9] Hall 140.

[10] See Geddes, *Cities in Evolution* 52.

[11] Lewis Mumford, "Who is Patrick Geddes?" *The Survey Graphic* 53 (1925): 524.

[12] Hall 148.

[13] Hall 174.

[14] Hall 146-150, and Geddes, *Cities in Evolution* particularly the chapter on "conurbations."

[15] Lewis Mumford, "The Fourth Migration," *The Survey Graphic* 54 (1925): 130.

[16] Stanley Buder, *Visionaries and Planners: the Garden City Movement and the Modern Community* (New York and Oxford: Oxford U P, 1990) chapter 7.
[17] Hall 89-135. For further analysis of these issues and influences, see Hall as a starting point.
[18] Hall 89-135.
[19] See Buder and Hall.
[20] Hall, *Cities of Tomorrow* 108. Welwyn quickly became a city of middle class commuters.
[21] See Richard Foglesong, *Planning the Capitalist City: The Colonial Era to the 1920s* (Princeton: Princeton U P, 1986) 186.
[22] Mumford, "Attacking the Housing Problem on Three Fronts," *The Nation* 99 (1919): 332-333.
[23] Edward K. Spann, *Designing Modern America: The Regional Planning Association of America and Its Members* (Columbus: Ohio State U P, 1996) 56. See also, Mumford, *The Golden Day: A Study in American Experience and Culture* (New York: Boni and Liveright, 1926) especially the chapter entitled "The Golden Day."
[24] Lewis Mumford, "Devastated Regions," *The American Mercury* III (1924): 218.
[25] Mumford, "The Fourth Migration" 130.
[26] Mumford, "The Fourth Migration" 130.
[27] Benton MacKaye, *The New Exploration* (1928; Urbana: U of Illinois P, 1962) 56.
[28] See Benton MacKaye, "A New England Recreation Plan," *From Geography to Geotechnics*, ed. Paul T. Bryant (Urbana: U of Illinois P, 1968) 161. This article originally appeared in *Journal of Forestry* XXVII, No. 8 (Dec. 1929): 927-930.
[29] Thomas 92-93 and the analysis of Mumford's *The Golden Day.*
[30] Mumford, "Regions – To Live In," *The Survey Graphic* 54 (May 1925): 151-152.
[31] "*The New Exploration* is a book that deserves a place on the same shelf that holds Henry Thoreau's *Walden* and George Perkins Marsh's *Man and Nature*; and like the first of these books, it has had to wait a whole generation to acquire the readers that would appreciate it." Mumford, introduction to Benton MacKaye, *The New Exploration: A Philosophy of Regional Planning* (1928; Urbana: U of Illinois P, 1962) vii.
[32] Mumford, introduction to MacKaye, *The New Exploration* viii. MacKaye biographical information comes from this Mumford introduction, Spann, *Designing Modern America*, and Paul T. Bryant, introduction to Benton MacKaye, *From Geography to Geotechnics* (Urbana: U of Illinois P, 1968).
[33] MacKaye, *The New Exploration* 69.
[34] MacKaye, *The New Exploration* 56.
[35] MacKaye, *The New Exploration* 56.
[36] Mumford, introduction to MacKaye, *The New Exploration* ix.
[37] Mumford, introduction to MacKaye, *The New Exploration* ix-x.
[38] Benton MacKaye, "Growth of a New Science," *From Geography to Geotechnics*, ed. Paul T. Bryant (Urbana: U of Illinois P, 1968) 22.
[39] MacKaye, "Growth of a New Science" 22.
[40] MacKaye, "Growth of a New Science" 26.
[41] Thomas, "Lewis Mumford, Benton MacKaye, and the Regional Vision" 71.
[42] Spann 22.
[43] See Paul Shriver Sutter, "Labor and Natural Resources: Colonization, the Appalachian Trail, and the Social Roots of Benton MacKaye's Wilderness Advocacy," presented at the University of Virginia History Department Seminar, September 28, 1998. Unpublished manuscript in possession of the author, 10-11.

[44] MacKaye, "Homesteads to Valley Authorities," *The Survey* 86 (1950), 496-498, reprinted in MacKaye, *From Geography to Geotechnics*, 35. See also, MacKaye, *Employment and Natural Resources* (Washington, DC: Government Printing Office, 1919).
[45] Benton MacKaye, "Some Social Aspects of Forest Management," *Journal of Forestry*, 16, 2 (February 1918): 210.
[46] MacKaye, "Some Social Aspects" 212.
[47] See MacKaye, *Employment and Natural Resources* 17-18; and see Spann, *Designing Modern America* 24.
[48] MacKaye, *The New Exploration* 67-69.
[49] Spann 26.
[50] Spann 27.
[51] Clarence Stein, introduction to MacKaye, "An Appalachian Trail" 326.
[52] See MacKaye, "An Appalachian Trail."
[53] MacKaye, "An Appalachian Trail" 327.
[54] MacKaye, "An Appalachian Trail" 328.
[55] MacKaye, "An Appalachian Trail" 328.
[56] MacKaye, "An Appalachian Trail" 328.
[57] MacKaye, "An Appalachian Trail" 329. See also MacKaye, *Employment and Natural Resources* 13.
[58] For the population decline and the social and economic changes in Virginia's Blue Ridge Mountains in this time period see Gene Wilhelm Jr., "Shenandoah Resettlements," *Pioneer America* 14: 15-41, and Robert Kyle, "The Dark Side of Skyline Drive," *Washington Post* 17 October 1993: C1.
[59] MacKaye, "An Appalachian Trail" 328.
[60] MacKaye, "An Appalachian Trail" 330.
[61] MacKaye, "An Appalachian Trail" 330.
[62] MacKaye, "An Appalachian Trail" 329.
[63] MacKaye, "An Appalachian Trail" 329.
[64] MacKaye, "An Appalachian Trail" 329.
[65] See Benton MacKaye, "Memorandum: Social Planning and Social Readjustment," 1921 Memorandum, MacKaye Family Papers, Dartmouth College Library, Hanover, NH.
[66] MacKaye, "Memorandum" 1.
[67] MacKaye, "Memorandum" 5-6.
[68] MacKaye, "Memorandum" 6.
[69] MacKaye, "Memorandum" 11-18.
[70] MacKaye, "An Appalachian Trail" 330.
[71] MacKaye, "Regional Planning and Social Readjustment" 18.
[72] Benton MacKaye, "The Great Appalachian Trail From New Hampshire to the Carolinas," *New York Times* (Feb. 18, 1923) 7:15.
[73] MacKaye, *Employment and Natural Resources*, 10.
[74] Lewis Mumford, Letter to Benton MacKaye, July 25, 1927, and Benton MacKaye to Lewis Mumford, July 30, 1927, MacKaye Papers, Dartmouth College Library, Hanover, NH.
[75] MacKaye, *The New Exploration* 200. See also the chapter entitled, "Controlling the Metropolitan Invasion," 168.
[76] MacKaye, *The New Exploration* 178ff.
[77] MacKaye, *The New Exploration* 179.
[78] Mumford, "Planning the Fourth Migration," *The Survey Graphic* 54 (1925): 130-133, and *The Golden Day*.

[79] MacKaye, "Tennessee – Seed of a National Plan," *The Survey Graphic*, 22 (1933): 251-254, 293-94, reprinted in *From Geography to Geotechnics*, and MacKaye, "Flankline vs. Skyline," *Appalachia* 20 (1934): 104-108.
[80] MacKaye, *Employment and Natural Resources*, 11.
[81] MacKaye, "The Great Appalachian Trail from New Hampshire to the Carolinas."
[82] MacKaye, *The New Exploration* for the use of the term "primeval." Also, see Carl Sussman, introduction, *Planning the Fourth Migration: The Neglected Vision of the Regional Planning Association of America*, ed. Carl Sussman (Cambridge: MIT P, 1976), for a discussion of the connection between the MacKaye Appalachian Trail article and the RPAA. And MacKaye, "An Appalachian Trail" 325-330.
[83] Benton MacKaye and Lewis Mumford, "Townless Highways for the Motorist: A Proposal for the Automobile Age," *Harper's Magazine* 16 (1931): 347-356.
[84] MacKaye and Mumford, "Townless Highways for the Motorist" 347-356.
[85] See MacKaye, "The Townless Highways," *The New Republic* 62 (1930): 93-95.
[86] Although roadside development was limited, suburban developers did benefit from the parkway's ability to get people into the suburbs, thus creating demand.
[87] MacKaye and Mumford, "Townless Highways for the Motorist" 352.
[88] MacKaye and Mumford, "Townless Highways for the Motorist" 350-356.
[89] MacKaye, "Tennessee – Seed of a National Plan."
[90] Sussman, introduction.
[91] Mumford, foreword to MacKaye, *The New Exploration.*
[92] MacKaye, "Appalachian Trail."
[93] MacKaye, "Flankline vs. Skyline" 104-108.
[94] Benton MacKaye, "Re Skyline Drives and the Appalachian Trail," MacKaye Family Papers, Dartmouth College Library, Hanover, NH.
[95] For a discussion on MacKaye and the Wilderness movement, see Paul S. Sutter, "Driven Wild: The Intellectual and Cultural Origins of Wilderness Advocacy During the Interwar Years (Aldo Leopold, Robert Sterling Yard, Benton MacKaye, Bob Marshall)," diss., U of Kansas, 1997.

Chapter 4

Metropolitan Boosters

The Parkway and Metropolitan Development

> The parkway serves a great variety of communication and recreation uses which cannot be obtained in ordinary highways. It provides special ways for passenger cars, which greatly facilitate the speed of these vehicles and relieve traffic congestion on those highways serving other forms of vehicular traffic. It is the best type of way for promoting a permanent residential development. It provides special paths for riding and walking and immediate access to open areas for picnicking, rest and play; although these are recreational features they are supplementary facilities to the larger ways of communication.[1]
>
> *The Graphic Regional Plan, 1929.*

1. INTRODUCTION

In the history of the regional parkway, an examination of the conflicting views of development in the region has been curiously omitted. Specifically, the opposing views of the regional visionaries (exemplified in the previous chapter by Benton MacKaye and Lewis Mumford) and those planners and business interests who believed in using the tools of city planning in the region – the metropolitanists – deserve wider consideration. This chapter discusses the metropolitanists – specifically the motivations and actions of the metropolitan planners, representatives of business interests, and municipal officials who aided in and benefited from the initial decentralization brought about through advances in transportation technology.

2. METROPOLITAN REALITY

The metropolitanists worked to rationally plan the decentralization of metropolitan areas. This rational approach provided the best structure for the economic development interests the metropolitanists represented and wished to benefit. The economic incentives to decentralize and the advances in transportation technology combined with the significant social, environmental, and urban reform roots of park and parkway development to coalesce around the suburban parkway. The motivations and actions of these metropolitanists *vis-a-vis* suburbanization ultimately served as a model for those who wanted the regional parkway to promote metropolitan development. Parkway boosters in both Virginia and Vermont wanted the perceived benefits of metropolitan planning in their rural/regional locales. The regional parkway boosters picked up on parts of the metropolitan model that seemingly fit their needs in the region. Unfortunately, as is discussed in subsequent chapters, the regional parkway boosters did not recognize the innate differences between their regional settings and the conditions that gave rise to, for instance, the Westchester County Park System. Finally, since concepts of wilderness and wilderness conservation, certainly not Benton MacKaye's later conception of wilderness, were not yet understood (conservation in the 1920s meant something different than it does now), the suburban park system served more than adequately as the model for regulated and at times constructed public space in the region. That is, the metropolitanists' vision of regional public land tended more toward Central Park, Van Cortlandt Park in the Bronx, and Saxon Woods in Westchester than it did toward MacKaye's primeval Appalachian range.

The substance of the metropolitanists' view is represented through three basic themes. First, the supporters of metropolitanization in the region drew upon the Bronx River Parkway and the other Westchester County parks and parkways as models. Opponents of the regional parkway in the wilderness, or, more specifically, opponents of the parkway as an alternative to the Appalachian Trail, viewed the Bronx River Parkway and the other Westchester parkways as an anathema – not to be built in the region. The second theme is suburban growth and the development of complementary economic growth, as represented by business-interest boosters intent on encouraging parkway development and public infrastructure investment as facilitators for economic development. These parkway boosters viewed the decentralizing nature of parkway development and the economic opportunities as positive economic incentives. Again, opponents of this type of metropolitanizing development viewed the region as a place to value rural economies, rather than the next logical place for metropolitan consumption. And the third theme, that of rational, ordered metropolitan growth, is represented by the Regional Plan of New York and Its Environs (RPNY & E). This plan accepted as a reality the idea that metropolitan growth could not be stopped, yet it could be ordered through rational planning. Supporters of the regional parkway, especially in Vermont, looked to incorporate, where possible, components of this rational planning model. The regional visionaries who opposed the metropolitanists did so on their own solid ideological foundations expressed in the previous chapter and articulated further in Lewis Mumford and Thomas Adams' debate over the merits of the 1929 RPNY & E.

2.1 Bronx River Parkway and the Westchester County Parks System

> Westchester County has not only improved its own living conditions but has set a new pattern for other regions.[2]
>
> Editorial Note, *American Planning and Civic Annual, 1934*

This statement, added at the end of an article by Stanley W. Abbott on the Westchester Parks system, well describes the attitude of the supporters of the regional parkway who believed in traditional metropolitan development. That is, the Westchester County model was thought to demonstrate the power and possibilities of rational planning tools used to promote economic development. The Westchester Parks system served as a direct model for Green Mountain Parkway boosters and as an indirect model for the Skyline Drive boosters. The construction of public infrastructure – parks and parkways – facilitated economic development, raised property values, allowed for efficient commuting, provided recreational outlets for citizens, and attempted to solve, at least in part, the problems of overcrowding and slums in New York City.[3] Except for the concern over solving urban overcrowding, the regional parkway supporters believed in the concepts and desired all the benefits of the Westchester County Park model.[4] The supporters of the regional parkways wanted to bring the tools of metropolitanization to the region.

In his dissertation on the history of the urban parkway, Glenn Orlin distinguishes between the earliest urban park roads and boulevards meant to serve urban parks and pleasure grounds in the latter part of the nineteenth century and early twentieth century, and the suburban parkways flowing radially outward from cities. Early twentieth century parkways were more comprehensive than their predecessors, and not merely extensions of parks.[5]

Orlin placed the Bronx River Parkway and the Westchester Parkways in this second era. They provided transportation corridors as well as recreational outlets. Further, using the tools of city planners, they facilitated suburban growth and the decentralization of the metropolitan area.

2.1.1 Bronx River Parkway Legacy

As discussed briefly in the second chapter, the Bronx River Parkway grew out of the need to resurrect the pollution-ridden Bronx River. The death of a number of water birds in a New York Zoological Society (Bronx Zoo) pond in the summer of 1904 seemed to be the seminal event in the establishment of the Bronx Parkway Commission and its efforts to clean up the river and build the Parkway.[6]

Although this environmental cleanup narrative dominated the history of the Bronx River Parkway, other scholars have begun to make the case that the construction of the Parkway had a broader rationale. In addition to public health, recreation, and transportation issues, the history of the Bronx River Parkway reveals further meaning in its reason for being. This broader history reveals other issues such as economics, modern planning processes, and the need for a public constituency for an expensive public project.[7]

In the last decade of the nineteenth century, as well as the first five years of the twentieth century, a number of attempts to clean up the Bronx River had been made. In 1905, the New York State Legislature established the Bronx Valley Sewer Commission in Westchester County. According to historian Marilyn E. Weigold, William W. Niles of the Board of Governors of the New York Zoological Society worked diligently for broader authority to deal with Bronx River pollution.[8] The New York State Legislature created the Bronx Parkway Commission in 1906, yet design and construction of the parkway did not begin until 1911. World War I interrupted construction for a few years, and, finally, in 1925 New York Governor Al Smith dedicated the Bronx River Parkway.[9]

The three early members of the Bronx Parkway Commission (BPC) were the innovators, motivators, and visionaries behind the Bronx River Parkway. Although the BPC hired engineers and landscape architects such as Jay Downer, William Thayer, L. G. Holleran, Herman Merkal, and Gilmore D. Clarke, the commissioners were primarily responsible for the Parkway idea.[10] The BPC included Madison Grant, a lawyer and businessman from Manhattan, William W. Niles, a lawyer and Democratic politician from the Bronx, and James Cannon, a real estate developer and banker from Scarsdale.[11] Each commissioner brought his own perspective and motivations to the efforts. Grant was a social reformer with an interest in conservation. He was also a eugenicist, scientific racist, and nativist. His enthusiasm for reforming unsanitary living conditions along the course of the Bronx River, a rhetorical theme consistently evident in the BPC's annual reports, underscores a greater meaning when understood within the context of his personal views.[12] Niles may have been the first to conceptualize the idea of the Bronx River Parkway. As Weigold explains, Niles visited Scotland in 1901 and was taken with the general cleanliness and recreational opportunities afforded by the River Ness in the midst of an urban setting.[13] Combining his long-standing interest in the Bronx Zoo, its viability, and his being from the Bronx itself, as well as his parks advocacy, Niles worked diligently to boost the Bronx River Parkway throughout his years with the BPC.[14] The third commissioner, James G. Cannon, filled the role of suburban development booster. As the owner of Scarsdale Estates, a land development company, Cannon donated some acreage to the BPC and consistently promoted efforts to boost the suburban development aspects of the project.[15] These three commissioners pushed the professional engineers and landscape architects to construct a Bronx River Parkway that reflected their conceptions of progressive reform, environmental reclamation, recreational opportunity, and suburban development.

The BPC and the professionals hired by the commission laid out the standards for the Bronx River Parkway project. In effect, these standards called for the primacy of the river in any design, a valuing of the natural environment, a clearing away of the built environment (especially dilapidated structures) – unless a well-made structure could be hidden through natural landscaping – and the use of local, natural materials whenever possible. Finally, these standards called for the road to follow the landscape and conform to the local geography throughout its course.[16] In addition to these design standards, the BPC recognized the pressures put on New York City by growth. The BPC viewed the Bronx River Parkway as at least part of the solution to this mounting pressure. In effect, the Parkway would help to guide growth out of the City.[17] In his address at the Ninth National Conference on City Planning, Downer described the suburbanization aims of the Parkway this way:

> The Parkway was not laid out on a comprehensive city plan but constitutes in itself a large item of planning, and provides a main axis, or backbone, for the development scheme of the important city and suburban territory which it serves.[18]

The construction of the Bronx River Parkway required the removal, through purchase or condemnation proceedings, of families, businesses, and other developments – in the name of cleaning up the Bronx River. In effect, these removals amounted to slum clearance. When combined with the other motivating factors for the entire project – the construction of a road used for recreation and commuting, the cleaning up of a polluted river, the creation of a park, and the introduction of infrastructure that would promote middle-class suburban development – it is evident that the Parkway served as a motivating model for reformers and boosters of economic development interests. The discussions of reform, progress, economic development, and the creation of parks and recreational outlets were certainly mirrored in the Westchester County Park System, and later by regional boosters and parkway supporters in the region.

2.1.2 Westchester County Parks and Parkway History and Legacy

The Westchester County Park System grew out of the Bronx River Parkway model to the point where many of the experts – engineers and landscape architects – who had previously worked on the Bronx River Parkway took on similar positions in Westchester. And, unlike the Bronx River Parkway project, the vision and implementation of the Westchester Park System emanated from these experts (rather than the commissioners). Both Gilmore Clarke and Thomas Adams pointed to Jay Downer and his leadership as the driving force in Westchester. [19]

Westchester County began to push for a comprehensive park and parkway system in 1922. As Weigold points out, the decade leading up to the early 1920s was one of rapid suburbanization of Westchester, especially the southern part. Typically, Westchester suburbanites had left New York City to take advantage of open space and recreational opportunities. Residents of New York City seeking recreational outlets also traveled to Westchester. Those who moved permanently to Westchester lived in low-density, single-family housing or apartment buildings situated in small towns and villages and relied on commuter transportation to get to and from their jobs in New York.[20] Westchester County sought legislative authority to manage the suburban pressures using the same tools available to the Bronx Parkway Commissioners. That is, the decision-makers in Westchester County consciously positioned themselves and county policy as facilitators of suburban growth. The decentralization of New York City brought people out into their county (to the country, really). The county welcomed this with the intent to apply efficient, comprehensive planning methods to manage the growth and create a rational and orderly metropolitan area. At the request of Westchester County, the New York State Legislature passed the "Westchester County Park Act" in 1923, establishing the Westchester County Park Commission. Between 1922 and 1945, the Westchester County Park Commission planned and developed valley stream parkways which flowed north from the Bronx, a transverse parkway, an extension to the Bronx River Parkway, and numerous parks and recreational outlets throughout the county.[21]

2.1.3 The Westchester County Park System and its Meaning

2.1.3.1 Recreation and Transportation

In a methodical, proactive planning process, the Westchester County Park Commission provided recreation and transportation opportunities for its citizens and helped to guide the growth and development of Westchester County. The Park Commission recognized the growth potential and development pressures brought about by the ongoing decentralization. As justification for their efforts, all they had to do was point to the degraded Bronx River Valley prior to the initiatives put forward by the Bronx Parkway Commission.[22] Public support coalesced around the Commission's efforts to provide recreation, prevent environmental nuisances, and create transportation corridors. The regional parks created in Westchester were "seen as a way to preserve and sometimes restore the natural beauty of a particular locale."[23] The creation, and oftentimes re-creation, of natural preserves and parklands within the region fit seamlessly into the suburbanization/regional-development paradigm. The Westchester County Park Commission took it upon itself to manage the suburbanization of Westchester through the construction of regulated park preserves adjacent to the suburban towns and villages that supplied New York City's middle-class workforce.

2.1.3.2 Economic Development – Land Valuation

The Westchester County Board of Supervisors, in looking to the Bronx River Parkway, proposed the Westchester County Park System partially because of the increased property values brought about through parkway development. While no explicit study[24] of the connection between parkway infrastructure and land values had been undertaken as of the mid-1920s, Gilmore Clarke, who was involved in both the Bronx River Parkway and the Westchester County Park System, claimed in a 1959 interview that economics and increased land values played a major role in justifying the development of the Westchester Parkways.[25]

John Nolen and Henry Hubbard's study, *Parkways and Land Values*, published in 1937 (though data collection and analysis had been done in the early 1930s), crystallized the known but previously unquantifiable notion that parkway development increased land values. Nolen and Hubbard highlighted the four primary benefits of parkway development. Businesses in a town whose customers and employees used the parkway benefited. Second, landowners who used the parkway to get to public parks benefited. Third, landowners who used the parkway for commuting witnessed increased value to their property. Finally, landowners who used the parkway itself for recreation also reaped the benefits of the parkway infrastructure.[26]

In their discussion of the Westchester Parkways, Nolen and Hubbard praised the geographic benefits of the New York City region. They noted that the parkways followed the valley streams, flowing northward out of New York City. These flows of traffic (really almost MacKaye-esque) naturally complemented the valleys and the course of the river. The construction of grade-separated crossings was also made easier by the valley stream location.[27]

The Nolen and Hubbard study attempted to attribute specific property-value increases to the Westchester Parkways. Methodologically this caused great difficulty. The development of the Bronx River Parkway, the Saw Mill River Parkway, and the Hutchinson River Parkway in southern Westchester made it difficult to find land outside of what the authors called "affected areas" – land not impacted by parkway development. Further, each of these three parkways followed a corridor that contained a commuter rail line,[28] which in itself altered the land values of the "affected area." Rather than attempting to dissect the influence on land values of each infrastructure, Nolen and Hubbard instead concluded:

> It would seem, therefore, that though the available information as to the Westchester County Parkways is immensely valuable qualitatively and can lead to very valuable quantitative convictions, on which we are justified in proceeding, still we cannot prove mathematically from this information, accurately in dollars and cents, the value of the parkways alone.[29]

The rail connection and the inability to separate its influence from the parkway is an important, yet seemingly overlooked piece of Nolen and Hubbard's study. The benefits of the Westchester Parkways emerged out of the concepts of recreation and transportation in a metropolitan region that was decentralizing from a still vibrant and previously established center. For those who wished to apply the Westchester model to a region without an established core employment and economic center, the commuter rail would of course be overlooked. As discussed in the next two chapters, metropolitan boosters in other locales attempted to apply the model to their region. And, while Hubbard did not publish the *Parkways and Land Values* study until 1937, Nolen certainly knew of the data and analysis he had completed in the early 1930s. As the supporters of the Green Mountain Parkway consulted with Nolen on the regional parkway and state planning in general, the connection between property values and the development of parkways and recreational opportunities in Westchester County became an important selling point.[30]

2.1.3.3 Metropolitan Planning Model

In 1934, Stanley Abbott, then Superintendent of the Blue Ridge Parkway, wrote of the Westchester County Park System begun in 1923:

> [. . .] Westchester is inherently a residential community. In its broadest aspects, the planning problem was to secure this suburban personality and direct the county's growth along lines consistent with its logical function in the greater metropolitan region. Following hard upon this fundamental was the requirement that the idealistic conception be combined with the intensely practical.[31]

The Westchester County Park System was, at its heart, a model for planning efficient metropolitan residential communities. The parkways combined with rail commuter transportation facilitated increased land values in compact, walkable village and town centers. Small town economies seemed to thrive. Amenities like parkway recreation and public park reservations further enhanced the economic development and rising property values, a prominent theme in the metropolitan parkway model. Finally, as Abbott pointed out, planning the Westchester County Park System dovetailed with the dominant economic development model; that is, the ideology of practical planning precluded the possibility of the economic or social reform espoused by the regional visionaries. In short, regional visionaries were not practical, he said, but metropolitanists were.[32] Further, when discussing the regional parkway, regional infrastructure, and planning in the region, the economic development boosters, planners, and parkway supporters looked for extant examples and dominant models. The Westchester County Park System clearly filled the void (as will be evident in the next two chapters).

The development of Westchester County served as an often-imitated yet never-quite-replicated model for the metropolitanists. The decentralization of New York City meant the metropolitanization of Westchester County. Ironically, the reality of Westchester actually justified the regionalists' view of the downside of metropolitanization, while simultaneously affirming the metropolitanists' positive view of metropolitan development. Again, on the downside, Westchester evolved into a suburb from its rural, small-town roots. On the upside, Westchester County is significantly more efficient in its use of land than most suburbs. Open space and recreational opportunities available to Westchester residents are superior to many, if not most, other suburban areas of North America. The combination of mass-transit commuter options and the parkways' limited access features were reminiscent of Benton MacKaye's writing on the "townless highway" concept. Conversely, Westchester County was quite rural before its suburbanization. Development clustered around the railroad stops along the three commuter lines. In the eyes of the metropolitanists, however, decentralization and the orderly creation of an efficiently planned suburban fabric provided an extremely positive model. Westchester's success thus motivated numerous other metropolitanizing efforts.

2.2 Business Boosters, Economic Development Impulses of Metropolitanism, and "See America First"

As a group, the metropolitanist supporters of the regional parkway consisted of a loose coalition of small-business groups, chambers of commerce, road and highway associations, and public and quasi-public commissions intent on bringing dollars, most often spent by tourists, to the region. Intent on increasing the economic and development potentials of their region, these groups employed the dominant themes of a rapidly changing culture. Road associations, for instance, promoted the liberating nature of the private automobile, yet their members were, often, also local businessmen interested in boosting the local economy. In their dual roles they realized that infrastructure development increased land values, economic activity and opportunity, and tax revenue. Also, as a corollary (that increased the auto-using constituency), the auto allowed the occupants to experience the landscape in a way previously unheard of. In addition to the rapidly passing scenery (also available on the railroad), the auto and the construction of roads allowed the user to tailor a recreational drive to specific interests and desires.[33] While the road associations may indeed have had altruistic visions of transporting every American into the "American landscape," economics certainly influenced the promotion of road construction.

The local boosterism received aid from the National Park Service (NPS) itself. Soon after the establishment of the NPS in 1916, Director Stephen T. Mather made a concerted effort to entice regional business, especially railroads, local chambers of commerce, road associations, and other business groups, to help promote the national parks, particularly those in the West. Basically, the NPS adopted a phrase already in use, "See America First," as a catchall concept to get Americans into the parks – preferably via the automobile.[34] This effort coincided with American involvement in World War I, closing off Europe to many travelers. Touring the West and, later, the East, was viewed as a method for increasing economic-development possibilities for a region while simultaneously preserving American landscape and history. (The American landscape and history had economic value in an auto-oriented tourist economy.)

In sum, the boosterism and promotion of regional economic development opportunities emerged as the coalescing point for a number of concepts. When examined separately and understood in detail, these concepts often seem contradictory. Ideas and themes centered upon boosterism and economic development were:

1. The automobile.
2. The need for automobile infrastructure.
3. The recognition that roads bring development.
4. The need for a consumer population to boost local economic prospects in the region.
5. National Park promotion.
6. Tourism – in the West, as an alternative to Europe.
7. Tourism – in the East after WWI.
8. Decentralization.
9. Improved transportation technology.
10. The decline of traditional local industries – due to changes in technology and changes in culture.
11. The rise of preservation and conservation – creating a new type of American landscape.
12. The blanket recognition that this new type of preserved landscape would promote development on the adjacent land – the food for consumerism.

As a whole, and when working in concert for the promotion of the regional parkway, for example, these concepts led down the path towards metropolitanist development.

2.3 Road Boosters

2.3.1 Westchester County and Nationally

Roadway boosters and highway associations pushed for the expansion of the highway infrastructure in metropolitan areas. These organizations worked to promote numerous projects on different geographic scales. Groups promoted national projects such as the Lincoln Highway through the early 1920s, regional projects such as the Eastern National Park-to-Park Highway Association in the late 1920s and early 1930s, and local or suburban projects like the proposed Hudson River Parkway (one of the proposed but never-built Westchester parkways).[35] Oftentimes members of local chambers of commerce also made up the membership of the highway promotion organizations – a daunting coalition of local advocates today and even more so during the 1920s.[36]

The Bronx Parkway Commission and the Westchester County Park Commission were boosters for economic development in their respective areas. The commissions, however, also facilitated development in a broader sense, rather than serving exclusively as advocates for specific roads and their specific routes. Local associations formed throughout Westchester and the New York metropolitan region to bring roads through certain towns and along certain routes and act as advocacy agents within the framework set up by the broader commissions. Two New York metropolitan-area parkway proposals supported by local groups exemplified this roadway boosterism. Weigold discusses a booster group intent on extending Riverside Drive north along the Hudson all the way to Peekskill. A second example came later in the 1930s in the form of the Merritt Parkway, the Connecticut extension of the Hutchinson River Parkway.[37]

A number of business associations, along with civic associations from Yonkers, pushed for continuation of Riverside Drive northward along the east side of the Hudson River in the latter half of the 1920s. Their influence led to discussion among the Westchester County Park Commission, representatives of these groups, and the mayor of Yonkers. The Commission then called upon Jay Downer to write a report outlining the feasibility of the project. Downer reported favorably on the prospect of the Hudson River Parkway, even though constructing the road through previously developed areas would raise its costs. Downer lauded the "great practical benefit [of the parkway] to the whole westerly side of the County."[38] The Hudson River Parkway ultimately proved too expensive to undertake as a project of the Westchester County Park Commission. The route would have crossed the adjacent railroad right-of-way a number of times. Additionally, pre-existing development would have required significant stretches of overpasses and viaducts, further increasing the costs. Weigold points out that in 1933 other business groups, notably the Lower Hudson River Association, requested reexamination of the proposed road, including a toll mechanism to offset costs. While a group of engineers did conduct a feasibility study, the Westchester County Park Commission did not build the road.[39]

The Hudson River Parkway is the most significant illustration of the role of local business and civic groups in the promotion of parkways. These groups recognized the benefits of parkways and other infrastructure that would facilitate economic development. The case of the Bronx River Parkway differed from the Westchester model in that the Bronx Parkway Commissioners advocated their own vision (for example Grant's conservation interests and Cannon's development interests). The Westchester County situation emerged as a project on a much broader scale, with the vision advanced less by the Commissioners than by careful consideration of individual projects by experts in the field such as Downer. This model allowed advocates of local economic development and municipal boosters to make their case to the experts and the decision-makers involved with the Westchester County Park Commission. The advocacy mirrored similar coalitions brought together around the country in support of road projects. The Westchester case was perhaps the best model for those boosting their respective regional parkways during the late 1920s and the 1930s.

The history of the Merritt Parkway demonstrates similar instances in which civic and business groups involved their own interests in the formulation of parkway development plans. The interested groups in Connecticut viewed Westchester as their model – or anti-model – just as Virginians and Vermonters did.[40] The case of the Merritt Parkway also demonstrates how the meaning, perceived benefits or detractions of the parkway, and design changed according to the involvement of local business and civic groups.

The Merritt Parkway grew out of the Westchester County Park Commission's earliest reports on the Hutchinson River Parkway. According to Merritt Parkway historian Bruce Radde, the Westchester County Park Commission had proposed extending the Hutchinson River Parkway into the southern Connecticut county of Fairfield. While the Westchester County Park Commission could not act on this proposal, no public agency in Connecticut picked up on this idea until late in the 1920s. In the meantime, groups organized on either side of the proposal.[41] Proponents, including the authors of the RPNY & E, wanted the parkway to deal with increased traffic congestion on US Highway 1, the Boston Post Road that ran through the communities along the New York and Connecticut shoreline. The Thomas Adams organization viewed a parkway connecting with the northern termination point of the Hutchinson River Parkway as a positive step towards mitigating the traffic problems in the New York Metropolitan area.[42] Opponents of the proposed road remained disinclined to support any infrastructure that would bring New Yorkers and suburban development to Connecticut's idyllic countryside.[43]

The initial supporters of the Connecticut extension to the Hutchinson River Parkway were primarily business organizations in the small towns along the Boston Post Road. These groups rallied behind the parkway in an unorganized fashion even though they had the support of a number of the southern Connecticut newspapers as early as 1923. In opposition to the Merritt Parkway (as originally proposed) stood the Fairfield County Planning Association (FCPA). This organization consisted of many "old Fairfield families, representing venerable New England propertied wealth and Republican virtues." Civic organizations with similar membership also lined up against the parkway.[44]

In his history of the Merritt Parkway, Radde explains that the FCPA supported a through road in Fairfield, yet only one that would promote higher property values, ruin the smallest number of wealthy estates, conform with the topography, and, once completed, blend in with the landscape as completely as possible. In an interesting and rather implausible claim, the

FCPA originally opposed the introduction of the parkway through Fairfield on the ground that they thought it would lower property values. Ultimately, the FCPA and its political supporters rallied around the parkway as long as it mirrored the design and engineering sensibilities of the Hutchinson River Parkway. Further, since the parkway would ultimately take more of an east-west course across many of the valleys and hills of southern Connecticut, the FCPA required significant design considerations so that the parkway would not scar the landscape.[45]

Through the help of Connecticut Congressman Schuyler Merritt, Connecticut secured federal funding in the early 1930s and began construction in 1934. Funding came from issuing state bonds and from the federal government. The Connecticut State Highway Commission designed, laid out the route, and constructed the Merritt Parkway, completing all sections by 1940. By designing it and keeping it in-house, the Connecticut Highway Commission bucked the trend of hiring technical help from the Bronx River Parkway/Westchester Park Commission cadre of experts.[46]

The importance of the Merritt Parkway as an exemplar for the metropolitanists is at least threefold. First, it demonstrated the power and effect of the Westchester Parkway. The Westchester County Park Commission intended to "Westchester-ize" southern Connecticut as early as 1923. Subsequently, Thomas Adams wrote so positively about Westchester's parks and parkways that he steadfastly endorsed its model throughout the region, Connecticut included. Second, the proponents and supporters of the Merritt Parkway argued for and against the project referencing Westchester. Less organized supporters viewed the parkway as a way to decrease congestion and promote land values. The better organized opponents viewed the extension of the Westchester parkways into their "backyard" as a way to lower property values, as a channel for bringing the riffraff of New York into the Connecticut countryside, and as a threat to their bucolic status quo. This sort of argument bordered on xenophobia – perhaps a logical extension of a regionalist ideology. Nevertheless, opposition to the initial proposal eventually resulted in greater local control over the parkway design, and the Merritt Parkway emerged as one of the best-designed, most-lauded examples of suburban parkways completed between World War I and World War II.[47]

2.3.2 "See America First" Connection

The "See America First" movement reflected local economic development boosterism and tourism on a regional scale, primarily in western states. The concept applied to the regional parkway and the metropolitanists in a more peripheral sense. Although See America First was primarily western in orientation, it did spread to the East at about the time of the initial push for eastern national parks. Further, the National Park Service, under its first director, Stephen T. Mather, did pick up the promotional benefits inherent in See America First. The popularity of See America First spread around the country to the point that proponents of regional parkways in Virginia and Vermont incorporated, if not the name itself, then a number of its boosterism objectives.

The "See America First" idea began as a plan to resuscitate the economies of western communities through tourism. Initial business boosters used the slogan to promote tourism, investment, and settlement in the West as early as 1905. By 1910, the Great Northern Railway began to use the phrase to promote travel on its rail lines. According to historian Marguerite S. Shaffer, over time it included "evolving ideas about tourism and commerce, scenery and history, the West and the nation."[48] Western boosters also used the idea behind See America First as a way to promote travel in the West once World War I had closed down European travel options for easterners. In addition to boosting the idea of economic opportunities, the leaders of See America First touted the benefits of the West's main resource – seemingly unlimited scenery. By boosting the "consumption" of this resource through tourism, and the inherent preservation of the scenery, the West would sidestep the pitfalls that befell the East: ruined natural resources and changing economies, for example.[49]

By the middle of the 1910s, two other organizations picked up on the See America First theme. The Panama-Pacific International Exposition of 1915 invoked the See America First slogan in advertising for the world's fair. Playing up the alternative to traveling to Europe during the war, the Exposition's organizers encouraged railroads to promote travel to the West Coast as tourism in and of itself. That is, the railroads could offer packages that took the tourist through Yellowstone National Park, the Grand Canyon, and Yosemite National Park while in transit to the Exposition. At about the same time, the Lincoln Highway Association, formed to promote the transcontinental Lincoln Highway, adopted See America First. The Lincoln Highway Association recognized the primacy of the automobile as a tourist and recreational vehicle. Shaffer notes that the automobile took the tourist into the landscape in a way that trains had not been able to,[50] marking a seminal change in the way Americans acted as tourists and as consumers of scenery, history, and the natural environment.

The National Park Service similarly picked up on the ideas associated with See America First. As Shaffer puts it, by embracing See America First, the National Park Service gave the phrase and its inherent meaning "official sanction of the national government and thus complet[ed] the transformation from western booster slogan to popular tourist emblem."[51] Coinciding with the planning of the Panama-Pacific International Exposition, Stephen T. Mather took a position in the Department of the Interior and combined some of the ideas associated with See America First with a plan to promote the national parks. Prior to this appointment, Mather had been a successful businessman in Chicago, championing business interests as a member of the Association of Commerce.[52] This business perspective pervaded the earliest attempts to build a constituency for the national parks. As Mather put it:

> Secretary [Franklin] Lane has asked me for a business administration. This I understand to mean an administration which shall develop to the highest possible degree of efficiency the resources of the national parks both for the pleasure and the profit of their owners, the people; the profit to be continually reinvested in the parks themselves. It is business to make these great public properties help themselves by adding to their yearly income provided by the Government; and it is business to make their common use by the people as cheap and as easy as possible. A hundred thousand people used the national parks last year. A million Americans should play in them this summer.[53]

Mather worked with Robert Sterling Yard, later a founder of the Wilderness Society, to promote the national parks as uniquely American and worthy of access by Americans. Most importantly, Mather worked to tie in the scenery available in national parks with economic development opportunities, tourism, local boosterism, and accessibility. Mather, by 1916 Director of the National Park Service, promoted the national parks to chambers of commerce, highway and road associations, auto associations, and wilderness groups. In addition to Mather's appearances and Yard's publicity work, Mather worked to spread the word about park development through publications such as *National Geographic* and *The Saturday Evening Post.*[54]

Mather's boosterism and the tie-in with economic development in the western parks spread to the East during this period. As is quite evident in the following chapters, the rhetoric associated with this early period of National Park Service development, spurred into action in part by the See America First concept, pushed local and regional booster groups to view the rational organization of the natural landscape (parks and open space) as tools to promote local business opportunities. Nearly all of the types of organization Mather addressed in the early years after the formation of the National Park Service worked to promote the economic development virtues of the regional parkway and Park Service involvement. Again, it is manifest that Mather's position was quite difficult. As director of a newly established agency charged with serving citizens and providing accessibility to citizens, he had to promote the benefits of the parks in any manner that would build a constituency. Playing on economic themes, as well as themes of identity associated with See America First, helped to establish and solidify this constituency. Unfortunately, in the eyes of many, including Robert Sterling Yard, the economic boosterism and development opportunities associated with national park promotion quickly overwhelmed its inherent purpose. By the mid 1930s, Yard joined Aldo Leopold, Benton MacKaye, Harvey Broome, and others to form the Wilderness Society to counter road development and the associated metropolitanization of the natural American landscape.[55]

The See America First concept helps to explain some of the additional connections between the development of tourism, the automobile, and the National Park Service (the primary conduit for the development of regional parkways) and the economic boosterism associated with the metropolitanization of the region. See America First and the role played by the Park Service in the early development of the national parks truly opens up an entire research project on its own.

3. THE REGIONAL PLAN OF NEW YORK AND ITS ENVIRONS AND THE PRO-METROPOLITAN GROWTH MODEL

The debate over putting the parkway in the region, especially along the Appalachian Mountain chain, pitted the regional visionaries against the metropolitan planners. The intersection of these two ideologies and the resulting conflict also erupted at other times in the history of planning in the United States. Both camps basically agreed on the malignancy of urban congestion and their concern for determining the best path for the imminent and rapidly occurring decentralization of cities and metropolitan areas. The great question each camp attempted to answer was, "How do we accommodate this decentralization?" The RPAA crowd had their say in the 1925 Regional Planning Number of the *Survey Graphic*, Henry Wright's *Report of the New York State Commission on Housing and Regional Planning*, Benton MacKaye's Appalachian Trail articles, several other publications by members of the RPAA, and the conference on Regionalism held at the University of Virginia in July 1931.[56] The metropolitanists of the RPNY & E published the eight volumes of *The Regional Survey of New York and Its Environs* in 1928 and *The Graphic Regional Plan* and *The Building of the City* in 1929. After the publication of the Plan, Lewis Mumford and Thomas Adams engaged in a public debate in the pages of *The New Republic* over the merits of metropolitan planning and the RPNY & E.[57] The debate demonstrated how wide the chasm between the metropolitanists and the regionalists was. This direct conflict of ideas expressed on paper mirrored the conflict that existed in theory and practice as the parkway developed in the region during the same period. The purpose of this section of this chapter is to discuss the theoretical and practical foundations of the RPNY & E and its relationship to the metropolitanist conception of the regional parkway.

3.1 Short History of the RPNY & E

The Committee on the Regional Plan of New York & Its Environs published the Regional Survey in 1928 and the Regional Plan in 1929. The Plan emerged after a number of other attempts at planning for the New York metropolitan area had failed to gain support for implementation earlier in the century. The other attempts included the New York City Improvement Commission Plan of 1907, the Plan of the Brooklyn Committee on City Plan of 1914, and the Port of New York Authority Plan of 1921 (begun 1917).[58] Charles Dyer Norton, formerly a colleague of Daniel Burnham in Chicago, became the initial leader of the Committee on the Regional Plan. Norton had intended to bring a Burnham-style planning process to the New York metropolitan area. Instead of attempting to enlist an organization along the lines of a Commercial Club of Chicago to fund the Regional Plan (such an organization really did not exist in New York, anyway), Norton solicited funding from the Russell Sage Foundation, which was already interested in progressive social issues.[59]

With Sage Foundation funding, the Committee on the Regional Plan began to put together its strategy. Joining Norton was fellow Chicagoan Frederic A. Delano. Norton and Delano brought to New York their Chicago roots, their experiences with the 1909 Plan of Chicago, and their desire to create a plan broad enough to establish its credibility within the constituent groups in the New York metropolitan area. In 1923, the Committee hired Thomas Adams,[60] an English city planner with roots in England's Garden City movement, as the General Director of Plans and Surveys. In addition, the Committee assembled an advisory group of professional planners, architects, landscape architects, and engineers that included John Nolen, Edward Bennett, Harland Bartholomew, George B. Ford, and Frederick Law Olmsted Jr. This promising group recommended, following Geddesian ideals, a survey prior to the plan.[61]

Discussions of the scope and purposes of the survey began in 1922. Surveys commenced by the end of 1923, and in 1928,[62] the Committee on the Regional Plan published eight books of survey data. In 1929, the Committee published two additional volumes entitled *The Graphic Regional*

Plan and *The Building of the City*. Soon after, the Committee on the Regional Plan changed its name to the Regional Plan Association (RPA) and published a second plan in 1968 and a third in 1996.[63]

3.1.1 Primary Ideas

During 1923, the members of the Committee on the Regional Plan's advisory group conducted a preliminary geographical survey of the region. Frederick Law Olmsted Jr., assisted by Henry V. Hubbard, surveyed the Long Island section. Thomas surveyed the two counties north of the Bronx, Westchester and Fairfield. John Nolen studied the Hudson River Valley, including the counties of Bergen, Rockland, Orange, and Putnam. Harland Bartholomew studied northern New Jersey, George B. Ford, assisted by E. P. Goodrich, analyzed central New Jersey, and Edward Bennett supervised a study of Staten Island and a section of New Jersey south of Manhattan.[64] From these reports, Adams prepared an outline and goals for the survey and the plan. This initial guide called for a regional zoning plan, agricultural zones for the provision of open space within growing areas, newly developed transportation opportunities to serve the region's circumference, rather than radially, and acquisition of additional public open space, especially near recreational opportunities. This initial outline included other issues related to industry development as well as the specific recentralization of certain industries and the dispersal of others. The point was, according to RPNY & E historian David A. Johnson, that Adams had created quite a comprehensive target for the entire project. Further, many of these initial concepts were ultimately incorporated into the survey and plan.[65]

The concepts put forward in this initial outline, survey, and synthesis led to the formation of survey teams that conducted the work which led to the 1928 publication of the eight-volume *Regional Survey of New York & Its Environs*.

The survey volumes led to the publication of *Graphic Regional Plan* and *The Building of the City* in 1929. The survey and plan volumes assumed the region's population would double by 1965. Many of the recommendations made in the RPNY & E incorporated methods to deal with this expected population growth. The plan also recognized the primacy of the small industrial manufacturing concern. According to Adams, the small firm drove the economy of the New York region. The small industries had to achieve some accommodation so long as the region wished to remain viable. Adams called for the movement of small industries outside of Manhattan and into the adjacent boroughs. This "diffused recentralization" would allow the small industries to take advantage of lower rents and future transportation improvements intended to accommodate decentralization in the region. The transportation infrastructure – rail, highway, and water improvements – were left primarily to the RPNY & E's transportation consultant, William J. Wilgus. Wilgus ("best known as the engineer who created the city's masterpiece of railroad engineering, the Grand Central Terminal"[66]) proposed two rail "beltways" to serve the region, allowing for Adams' "diffused recentralization."[67] Wilgus also called for transit improvements, road and boulevard construction, and parkway construction to further accommodate the projected population growth in the region and the complementary industrial recentralization.[68]

The Regional Plan of New York & Its Environs provided a blueprint for the accommodation of growth in the metropolitan area. The authors of the RPNY & E intended to improve, through rational planning, transportation infrastructure, land use patterns, zoning, and the provision of regulated open space in order to facilitate the growth of population and industrial development within the metropolitan area. Although the Committee on the Regional Plan conducted regional surveys, the plans promoted by the Committee promoted progress through the metropolitanization of the region. The surveys did not necessarily recognize the uniqueness of the different parts of the region and call for appropriate planning that valued uniqueness.

3.1.2 References to Parkways

During the early surveying stage of the Regional Plan of New York & Its Environs, Thomas Adams took the opportunity to address the National Conference on City Planning held in New York in 1927. In his address entitled, "Regional Highways and Parkways in Relation to Regional Parks," Adams applauded the work of the Westchester Park Commission and the development of the Westchester parkways to that point. Regarding Westchester, he said:

> In preparing the Regional Plan for New York we do not need to put on our plan any higher standard of open space for the remainder of the region than is already provided in the county of Westchester. In other words, all that the regional planner has to do is try and get the other counties in the region to follow that example. We do not have to ask people to do anything that has not already been done and proved to be profitable.[69]

To the Director of Surveys and Plans for the Regional Plan of New York & Its Environs, Westchester's park and parkway system set the standard for his conception of metropolitan/regional planning. Adams wanted the model duplicated throughout the New York metropolitan region. The Westchester model provided profitability to the municipality, regulated open space for the community, created recreational opportunities for the homeowner, and provided an efficient pathway for the commuter. Westchester offered all of this in a space that would, by 1965, include its share of an expected 21 million people.

By 1929 and the publication of *The Graphic Regional Plan*, Adams had incorporated the 1927 presentation into the initial volume of the Regional Plan. *The Graphic Regional Plan*'s section on parkways and highways in the chapter entitled "The Regional Highway System" outlined the proposed parkway system for the entire region. Again applauding the Westchester parkways, Adams used the existing parkways as a foundation for those in Long Island, Connecticut, and New Jersey. *The Graphic Regional Plan* called for belt parkways to complement the arterial parkways and provide for more fluid transportation corridors throughout the region.[70]

Towards the end of the section entitled "The Regional Highway System," Adams explained at some length the economy of constructing parkways in the metropolitan region. The recreational component of the parkway greatly increased surrounding property values and therefore "creat[ed] revenues" needed to cover the initial construction costs.[71] The recreational and open space components of the parkway further enhanced it as a tool for metropolitan planning. Adams argued that the parkway's value as a commuter road was heightened by the segregation of vehicle types. Prohibiting trucks on parkways added value to nearby residential neighborhoods. In and of themselves, the open spaces, of course, helped to raise adjacent property values. By routing the parkway through recreational areas, consumers of these recreational opportunities had alternatives to driving. Finally, Adams argued that it was most appropriate for the parkway to "run through parks,"[72] thereby fusing the concept of the parkway and the park. The RPNY & E, Adams, and *The Graphic Regional Plan* validated this idea that the parkway and the recreational area belonged together in metropolitan planning. As others looked to the New York region, to Westchester, they perceived this validation. As the professionals who worked for the Committee on the Regional Plan and the Westchester Park Commission took their expertise to other places, they took with them the positive examples built and planned in the New York metropolitan region. However, those professionals did not recognize the uniqueness of Westchester and the New York metro area.

3.2 Adams/Mumford Debate

Lewis Mumford reacted harshly to the publication of the Regional Plan of New York & Its Environs. In a series of critical remarks that, for all intents and purposes, had begun to brew as early as 1925, Mumford shredded the RPNY & E's arguments, beginning with the basic premise. The debate between Mumford and Adams raged in the pages of *The New Republic*, while the basic ideas of the argument played out as the same conflict shaped the regional parkway in the decade after the publication of the work of the Committee on the Regional Plan.

Perhaps the most important issue with regard to this debate and the development of the regional parkway is the assumption by the RPNY & E planners that they could plan for significant population growth. Essentially, the planners were saying they could make the transportation infrastructure accommodate a doubling in the size of the region's population even though the central city was already congested. Moreover, the concept that growth was inevitable removed any question over population growth from public scrutiny.

A staff writer for *The Survey* wrote a dispatch on the International Town, City and Regional Planning Conference held in New York in mid-April 1925. The conference coincided with *The Survey*'s publication of the "Regional Planning Number" that same month. By tying in some comments made by President Calvin Coolidge a few days after the conference, the staff writer succinctly summed up the path the Adams/Mumford debate would take after the publication of the work of the Committee on the Regional Plan.[73]

> The President was careful not to 'presume to judge between those investigators who conclude that the cities must inevitably go on with their rapid rate of growth and those others who tell us that the transportation and industrial program must be made to counteract this tendency and bring diffusion of the population masses.' With him on the fence will be found today, perhaps, the great majority of American planners.[74]

By 1929, Lewis Mumford no doubt viewed Thomas Adams as more than a fence sitter. Adams, according to Mumford, had accommodated the spread of the "diluted form" of the city. RPNY & E historian David A. Johnson wrote of Adams being "deeply hurt" after Mumford cited the Committee on the Regional Plan for one of *The New Republic*'s "Booby Prizes for 1929."[75] Adams met Mumford's initial criticism of the work of the RPNY & E with considerable dismay, but did try to reconcile the differences between the two. Unfortunately for this attempt at reconciliation, Adams seemed to challenge Mumford to better study the work of the Committee on the Regional Plan.[76] The result was, of course, the well-known exchange between Mumford and Adams published in *The New Republic* during the summer of 1932.[77]

In the published debate, Mumford criticized nearly the entire RPNY & E effort, employing his stinging rhetoric to ridicule recommendations of the Regional Plan and its authors. Adams defended the work of the RPNY & E and criticized Mumford's refusal to accept political and economic realities. The gap between the regional visionaries and the metropolitan planners came down to the primary premise of the Regional Plan. Mumford refused to believe the New York metropolitan area had to accept the population growth called for in the RPNY & E analysis. To Adams, growth was inevitable and he intended to plan for it. Mumford believed it was inevitable only if the regional planners planned for it. Further, planners could plan against it. Benton MacKaye chimed in with his own criticism of the Committee on the Regional Plan. Comparing the RPNY & E plan to making something painful a little less so, MacKaye wrote:

> Why make the way 'swifter and less painful'? Why not slower and more painful? Why not deliberately plan to achieve something that looks nothing like an asset and something like a liability [. . .]. The more subways the more suburbs, the more suburbs the more subways, *ad infinitum* [. . .]. The old story, as *The Survey Graphic* has said: moving 'people each day from places where they would rather not live to places where they would rather not work, and back again.'[78]

In the region that did not have the congestion pressures inherent in the New York metropolitan area and where increased population was desirable, the development of transportation infrastructure modeled on the Regional Plan painted a disheartening picture for those regional visionaries who believed in the uniqueness and the quality of the region – a daunting picture for those who believed the "hope of the city lies outside of itself."[79]

4. CONCLUSION

The concept of the metropolitan reality, outlined in this chapter, provided the model for the proponents of the parkway in the region. These parkway boosters viewed the parkway as a way to increase economic development opportunities along the lines of the metropolitan model. A number of ideas, events, and issues rapidly worked into American life during the first three decades of this century. The metropolitan model emerged from a number of converging issues. The rise of the automobile and the suburban parkway served as a catalyst for an economy fueled by increasing property values and consumerism. The automobile also opened the regional landscape and helped to democratize the national parks. Finally, planners advanced the metropolitan model through the belief that they could efficiently plan for growth, congestion, and decentralization all at once. The next two chapters will show that many of the regional parkway proponents used one or more of these arguments in their promotion of the regional parkway. While a direct link to the metropolitanists does not always exist, ideological influences are quite evident.

---

[1] Committee on the Regional Plan of New York and Its Environs, *The Graphic Regional Plan*, Vol. 1 (New York: Committee on the Regional Plan of New York and Its Environs, 1929) 269-271.

[2] See the editorial note accompanying Stanley Abbott, "Ten Years of the Westchester County Park System," *American Planning and Civic Annual* 5 (1934): 125-126.

[3] See Ken Jackson, *Crabgrass Frontier* (Oxford: Oxford U P, 1985) 174-175. Jackson noted that the suburban boom of the 1920s, aided by the automobile, met with the approval of urban "housing reformers" concerned with overcrowding and slums.

[4] This concept is evident in the case study chapters.

[5] See Glenn Orlin, "The Evolution of the American Urban Parkway," diss., George Washington U, 1992, 103.

[6] Jay Downer, "The Bronx River Parkway," *The Proceedings of the Ninth National Conference on City Planning* (Kansas City: n.p., 1917) 91.

[7] Randy Mason, "Memory Infrastructure: Preservation, 'Improvement' and Landscape in New York City, 1898-1925," diss., Columbia U, 1999, Chapter 6, "Bronx River Parkway: The Nature of Improvement." Environmental clean up dominates the historiography of the Bronx River Parkway.

[8] Marilyn E. Weigold, "Pioneering in Parks and Parkways: Westchester County, New York, 1895-1945," *Essays in Public Works History* 9 (February 1980): 3-4. Other early history of the Bronx River Parkway comes from Ethan Carr, "The Parkway in New York City," *Parkways: Past, Present, and Future* (Boone, NC: Appalachian Consortium P, 1987) 121-128, and Domenico Annese, "The Impact of Parkways on Development in Westchester County, New York City, and the Metropolitan New York Region," *Parkways: Past, Present, and Future* (Boone, NC: Appalachian Consortium P, 1987) 117-121.

[9] Weigold 10.

[10] "Reminiscences of Gilmore David Clarke: Oral History" (Oral History Collection of Columbia University, 1959) Chapter 2. Hereafter Clarke, COHC.

[11] See Clarke, COHC, Chapter 2, Weigold 4, and Mason Chapter 6. The initial BPC was established in 1906 and included other commissioners, D.H. Morris from the Bronx and J. Warren Thayer, an engineer from Scarsdale. In 1907, the BPC was reorganized into the three-person commission consisting of Grant, Cannon, and Niles. It is this commission that Clarke referred to when discussing the visionaries behind the BRP.

[12] Mason Chapter 6, for a detailed discussion.

[13] Weigold 35, note 6, and Orlin, "The Evolution of the American Urban Parkway" 146.

[14] Weigold 3-4, 35, note 6. Also Mason Chapter 6.

[15] Mason Chapter 6, Weigold 4 and 35, note 7.

[16] Bronx Parkway Commission, *Final Report of the Bronx Parkway Commission* (Bronxville: J.B. Lyon, 1925) 69.

[17] Bronx Parkway Commission, *Annual Report of the Bronx Parkway Commission* (Bronxville: J.B. Lyon, 1912) 15.

[18] Downer 91.

[19] See Clarke, COHC Chapter 31 and Thomas Adams, "Regional Highways and Parkways in Relation to Regional Parks," *The Proceedings of the Nineteenth National Conference on City Planning* (Philadelphia: Wm. F. Fell, 1927) 178.

[20] Weigold 11. The electrification of the commuter rail lines and the construction of Grand Central Terminal made commuting to and from Manhattan an easier prospect for people settled in Westchester.

[21] Weigold 1-12 for the date of the establishment of the Westchester County Park Commission and its development plan.

[22] Weigold 11.

[23] Weigold 12.

[24] No study the size of John Nolen or Henry Hubbard's study published in 1937. Nolen and Hubbard used data available up through 1931 for the 1937 study. In fact, Nolen and Hubbard began their analysis in 1930, but were unsatisfied with the work. It was not until Nolen's death in 1937 that Hubbard completed and published the work. See John Nolen and Henry Hubbard, *Harvard City Planning Studies, Volume XI: Parkways and Land Values* (London: Oxford U P: 1937) introduction.

[25] See Clarke, COHC Chapter 3, and Nolen and Hubbard 6.

[26] Nolen and Hubbard 4.

[27] Nolen and Hubbard 72. The valley stream location also made these roads susceptible to constant flooding during rainstorms – a distinct childhood memory of the author.

[28] Nolen and Hubbard 98-100. Nolen and Hubbard wrote that a study of the Hutchinson River Parkway separate from the influence of the New Haven Line commuter railroad may have been possible, but they did not undertake this.

[29] Nolen and Hubbard 100.

[30] See Chapter 6 for this discussion.

[31] Abbott 125-126.

[32] I believe that it is pretty clear that Abbott's last sentence of the previous quote, "[. . .] was the requirement that the idealistic conception be combined with the intensely practical," was a direct statement in support of Thomas Adams and the RPNY & E, and a slap at Mumford, MacKaye, and the RPAA. Gilmore Clarke noted that Adams had been close to Jay Downer during the development of the Westchester County Park System and had often aided Downer's team. See Clarke, COHC Chapter 3.

[33] Marguerite Shaffer, "Negotiating National Identity: Western Tourism and 'See America First,'" *Reopening the American West*, ed. Hal K. Rothman (Tucson: U of Arizona P, 1998) 139.

[34] Shaffer 139-143. Shaffer explores the concept of 'See America First' in great detail.

[35] See M. Christine Boyer, *Dreaming the Rational City: The Myth of American City Planning* (Cambridge, MA: MIT P, 1983) 178-179, Weigold 23; "Thinks Road Will Boost Tourist Travel: Thatcher Comes to Confer On Asheville Link in Route," *Asheville Citizen* 26 November, 1931.

[36] This connection is quite clear in the subsequent chapters.

[37] See Weigold 23, and Bruce Radde, *The Merritt Parkway* (New Haven and London: Yale U P, 1993).

[38] See Westchester County Park Commission, *Annual Report* (White Plains, NY: Westchester County Park Commission, 1927) 20, and Weigold 24.

[39] See Weigold 25.

[40] See chapters 5 and 6 for the detailed discussion.

[41] See Radde 14-15.

[42] Committee on the Regional Plan of New York and Its Environs, *The Graphic Regional Plan*, Vol. 1, 273.
[43] Radde 14.
[44] Radde 14
[45] Radde 16.
[46] See Radde 19-21.
[47] Radde 10-30. The history of the Merritt Parkway also shows a legacy of corruption throughout its construction.
[48] Shaffer 123.
[49] Shaffer 126-127.
[50] Shaffer 137.
[51] Shaffer 137.
[52] Enos A. Mills, "Warden of the Nation's Mountain Scenery," *The American Review of Reviews* 51 (1915): 428.
[53] Stephen T. Mather, "The National Parks on a Business Basis," *The American Review of Reviews* 51 (1915): 429.
[54] Shaffer 142-143, and Donald C. Swain, *Wilderness Defender: Horace M. Albright and Conservation* (Chicago and London: U of Chicago P, 1970) 44-48.
[55] Paul S. Sutter, "Driven Wild: The Intellectual and Cultural Origins of Wilderness Advocacy During the Interwar Years (Aldo Leopold, Robert Sterling Yard, Benton MacKaye, Bob Marshall)," diss., U of Kansas, 1997.
[56] For a discussion on the Regionalism conference at UVA, see Edward K. Spann, *Designing Modern America: The Regional Planning Association of America and Its Members* (Columbus: Ohio State U P, 1996) 126-129.
[57] See Carl Sussman, *Planning the Fourth Migration: The Neglected Vision of the Regional Planning Association of America* (Cambridge: MIT P, 1976) 221.
[58] See David A. Johnson, "Regional Planning for the Great Metropolis: New York Between the World Wars," *Two Centuries of American Planning*, ed. Daniel Schaffer (Baltimore: Johns Hopkins U P, 1988) 167-174.
[59] See Robert Fishman, "The Regional Plan and Transformation of the Industrial Metropolis," *The Landscape of Modernity: Essays on New York City, 1900 – 1940*, eds. David Ward and Olivier Zunz (New York: Russell Sage Foundation, 1992) 110-111. The Russell Sage Foundation had provided support for a social survey of Pittsburgh and the planning of Forest Hill Gardens prior to supporting the Regional Plan of New York & Its Environs.
[60] For a complete history of the early organization of the Committee on the Regional Plan and the entire Regional Plan of New York & Its Environs, see David A. Johnson, *Planning the Great Metropolis: The 1929 Regional Plan of New York and Its Environs* (London: E & F N Spon, 1996) 80ff.
[61] See Fishman 111-112; Johnson, "Regional Planning for the Great Metropolis," 175-177; Committee on the Regional Plan, *Graphic Regional Plan* introduction.
[62] Johnson, *Planning the Great Metropolis* 92-95.
[63] See Regional Plan Association, *A Region at Risk: The Third Regional Plan for the New York-New Jersey-Connecticut Metropolitan Area* (Washington, DC: Island P, 1996) 1-2.
[64] See Johnson, *Planning the Great Metropolis* 96-105.
[65] Johnson, "Regional Planning for the Great Metropolis" 177, and Johnson, *Planning the Great Metropolis* 104-108 (for a more thorough analysis).
[66] Fishman 112.

---

[67] Fishman 112-113, and Committee on the Regional Plan, *Regional Survey of New York and Its Environs: Transit and Transportation, and a Study of Port and Industrial Areas and Their Relation to Transportation*, Vol. IV, by Harold M. Lewis, William J. Wilgus, and Daniel M. Turner (New York: Regional Plan of New York and its Environs, 1928) 164-167. The term "diffused recentralization" comes from *The Graphic Regional Plan*, 150.

[68] See the section authored by Wilgus in the Regional Survey, "Transportation in the New York Region," *Regional Survey* Vol. IV, 161-188.

[69] Adams, "Regional Highways" 179.

[70] Committee on the Regional Plan, *The Graphic Survey* 272-285.

[71] Committee on the Regional Plan, *The Graphic Survey* 272-285.

[72] Committee on the Regional Plan, *The Graphic Survey* 272-285.

[73] "Bigger and Better Cities?" *The Survey* 54 (1925) 216.

[74] "Bigger and Better Cities?"

[75] See Johnson, "Regional Planning for the Great American Metropolis" 179.

[76] Johnson quotes Adams' response to Mumford after Mumford published his initial criticism in *The New Republic*. See Johnson, "Regional Planning for the Great American Metropolis" 179.

[77] Thomas Adams, "A Communication in Defense of the Regional Plan," *The New Republic* 71 (1932): 207-210, Lewis Mumford, "The Plan of New York," *The New Republic* 71 (1932): 121-126, Mumford, "The Plan of New York: II," *The New Republic* 71 (1932): 146-154. See also Sussman for reprinted versions of this debate.

[78] Benton MacKaye, "New York a National Peril," *Saturday Review of Literature* 23 August, 1930, 68. MacKaye reviewed R.L. Duffus, *Mastering a Metropolis: Planning the Future of the New York Region* (New York: Harper & Brothers, 1930). According to Johnson, this publication was the "popularized" version of the Regional Plan.

[79] This quote is from Mumford's article, "Regions – To Live In," *The Survey Graphic* 54 (1925): 151.

Chapter 5

The Skyline Drive

Visionaries and Boosters – 1

> The greatest single feature, however, is a possible sky-line drive along the mountain top, following a continuous ridge and looking down westerly on the Shenandoah Valley, from 2,500 to 3,500 feet below, and also commanding a view of the Piedmont Plain stretching easterly to the Washington Monument, which landmark of our National Capital may be seen on a clear day. Few scenic drives in the world could surpass it.[1]
>
> *Southern Appalachian National Park Commission, 1924.*

> The critical American wilderness is at present jeopardized as it never was before. I refer immediately to slashing open of our eastern wilderness belts by skyline roads, specifically by the proposed parkways for the White and Green Mountain Ranges in New England and the Park-to-Park Highway in the Southern Appalachians [...].[2]
>
> *Benton MacKaye, 1934.*

1. INTRODUCTION

The Shenandoah National Park's Skyline Drive developed out of disparate visions of planning in the region. This conflict between the regionalists and the metropolitanists shaped the development of the Skyline Drive through its early history and continues to influence the idea of the Drive today. As a parkway in the region, the Skyline Drive did not emerge directly out of the earlier parkways (as the next generation or the next scale of parkway). Neither the pursuit of urban reform, environmental reform, commuter travel, nor suburban recreation alone spurred on its development.

Moreover, the Drive did not develop solely as a product of planning in the region promoted by the metropolitanists. Instead, the Drive was shaped by the conflict between the regionalists and the metropolitanists. The regional planning theorists subscribed to an ideology that privileged the small town, the local industry, and the pedestrian traveler. The metropolitanists championed the automobile and economic development through boosterism, and applied the tools of city planning to the region. The struggle to build the Appalachian Trail, "a project in regional planning," according to Benton MacKaye,[3] at the same time and in the same physical location as the Skyline Drive, demonstrated the dispute between the regional visionaries and the metropolitanists.

This chapter discusses the planning and development of the Skyline Drive and the inherent conflicts that influenced its ultimate shape, literally and philosophically. Following a brief summary of the Drive's history, this chapter examines in-depth the primary players: supporters of the Shenandoah National Park and Skyline Drive who looked to the metropolitanists as their guide; the Potomac Appalachian Trail Club membership, their inspiration, and their own clashes over the Drive; Benton MacKaye and regional planning; and the indigenous community of soon-to-be-displaced Virginians along the Blue Ridge. The rhetoric and actions of each of these groups demonstrated their conflicting visions.

Today, both the Skyline Drive and the Appalachian Trail exist within the boundaries of the Shenandoah National Park and visitors to the Park experience the outcome of this conflict. Today's experience along the Skyline Drive is much as Benton MacKaye predicted it would be. Each fall tens of thousands of Shenandoah National Park visitors crawl along the Skyline Drive in massive traffic jams looking at the foliage. They get the broad brushstroke of the landscape, yet miss the specific details of the landscape – often just a blur from the car window.

2. SETTING UP THE CONFLICT

The major players in the development of the Shenandoah National Park, the Skyline Drive, and the Virginia section of the Appalachian Trail helped determine the form of the Skyline Drive. The Potomac Appalachian Trail Club members, who were carrying out Benton MacKaye's plan for the Trail, represented the regional visionaries.[4] The metropolitanists, on the other hand, are best characterized as boosters and were interested in regional economic development and regulated recreational opportunities. In Virginia, the metropolitanists were members of private and quasi-public statewide and local development groups interested in the benefits of a nationally recognized tourist attraction, as well as various professionals in the National Park Service and the Department of the Interior. As the prime motivator for the construction of the Skyline Drive, this group promoted the Shenandoah National Park and Skyline Drive as means of access to the national parks in the East, recreation through automobile use, local and regional economic development, and the ability to build a national constituency for the park concept.

A third group, the Virginians displaced by the Shenandoah National Park and the Skyline Drive, exist today as a painful reminder of the human element of this planning controversy. This group perhaps had the most to gain through idealism and social reform associated with the regional planning ideology of MacKaye and Mumford, which privileged the small town, the cultural uniqueness of regional culture, and the economy outside the dominating forces of late capitalism. In contrast, the metropolitanists believed in the worthiness of their actions – the across-the-board faith in the value of a tourist industry, automobile consumerism, and metropolitan development – to the extent that displacing a relatively small group of families not in the mainstream of society was, to them, of little consequence.

2.1 Regional Vision Players

The Potomac Appalachian Trail Club (PATC) formed in 1927 to construct the Appalachian Trail in a place accessible to the Washington, DC area. The PATC grew out of the Appalachian Trail Conference, the umbrella organization formed in an effort to organize and execute Benton MacKaye's 1921 plan for the Appalachian Trail.[5] The PATC members planned to create the Appalachian Trail according to MacKaye's writings:

> And this is the job that we propose: a project to develop opportunities – for recreation, recuperation, and employment – in the region of the Appalachian skyline.[6]

In their attempt to implement the vision, the PATC membership foresaw the impact the Skyline Drive would have upon the Appalachian Trail. Harvey Broome, an early member of the Appalachian Trail Conference and later a founding member of the Wilderness Society, wrote in *The Living Wilderness*:

> Obviously, if automobiles with their urbanizing influence were to follow the Trail, the purpose for which it was conceived, namely, to offer a medium of offsetting and balancing industrialized urban life, would be destroyed. The Skyline Drive was no isolated case. There was strong agitation for skyways in Vermont, in the South, and elsewhere.[7]

So practically from the start of the planning of the construction of the Appalachian Trail in the Blue Ridge Mountains of Virginia, MacKaye's vision and philosophy regarding the Trail went against the construction of the Skyline Drive. The PATC membership itself struggled with this disagreement in philosophy and vision, as discussed later in this chapter.

2.2 Metropolitan Reality Players

The Southern Appalachian Park Commission endorsed the possibility of a "skyline" drive in its 1924 report that recommended the formation of the Shenandoah National Park. This concept was appropriate given the rise of the automobile as a recreational tool. Consider the words of landscape architect Charles Elliot 2nd in 1922:

> The speed of automobiles has made a great expanse of open country accessible to the automobile owner. The whole countryside has become the motorist's park. It would seem fair under these conditions that the parks in the city should be designed primarily for the use of those to whom this open country is not accessible and that the pedestrian should have prior consideration.[8]

In Virginia, members of the development and tourism communities pushed for the construction of the Skyline Drive and the complete access of the automobile. The Virginia State Commission on Conservation and Development and its chairman, William E. Carson, represented these groups, as well as the landscape architects and administrative officials in the National Park Service and the Department of the Interior. Moreover, the development and tourism community formed its own groups throughout Virginia to push for the formation of the Shenandoah National Park and the Skyline Drive. Ferdinand Zerkel, a longtime supporter of the park, summed up the motivations of these groups in 1931:

> The State of Virginia became interested in the Park largely on commercial prospects for all the state and the Skyline Drive was to bring the folks as tourists for scattering over all the state.[9]

The boosters' argument was based on the need to build a broad constituency for the Shenandoah National Park – an effort that was eventually bolstered by the possibility of the Skyline Drive. Additionally, the rise of automobile use and its acceptance as a form of recreation in and of itself aided the argument. Finally, the dominant theories of the landscape architecture and planning professions during this time considered the regional parkway the appropriate model for experiencing the landscape outside the city.

2.3 Displaced Communities of Virginia's Blue Ridge Mountains

Establishing the Shenandoah National Park and the Skyline Drive resulted in the displacement of approximately 500 families who prior to 1936 had been living within what has now become the park boundary. The number was (and still is) disputed. According to Charles Perdue and Nancy J. Martin-Perdue, researchers who have long studied the history of the mountain families, federal officials arrived at the number 500 merely by counting houses within the proposed boundaries in 1934. But, as the Perdues note, news of the proposed establishment of the park was publicized as early as 1924. During the years 1924 to 1934, a number of families left the mountain area in anticipation of the displacement.[10] So, the estimate of 500 families may be only half of the actual number displaced by the park.[11]

In addition to the changes brought about by the Shenandoah National Park and the Skyline Drive, prior socioeconomic changes had disrupted the community and culture of the "mountain folk." In his study on the resettlement of the people displaced by the Park, Gene Wilhelm notes that changes in industry, technology, and culture caused a general shift in demographics. The changes in the local economy and culture began soon after the end of the Civil War. By the early 20th century, over-timbering and the depletion of iron-ore deposits brought on even more rapid changes. New advances in textile technologies ended the local mountain tanbark industry, and the chestnut blight during the first twenty years of the century drastically diminished the local mountain economy.[12]

Prior to the Shenandoah displacements and the manner by which people were moved off their land by the state of Virginia and the United States government, there existed a unique population upon whose lives the disparate ideologies of regional planning were imposed. The regionalist ideology valued this culture, whereas the metropolitanist ideology could not. Charles Perdue and Nancy J. Martin-Perdue perhaps unknowingly commented on this conflict in a 1991 article. They wrote of an English folklorist named Cecil Sharp and his trip through Appalachia in 1917. Sharp's concern for the traditional culture in the face of economic expansion and industrialization and the recognition of this concern by the Perdues is really an acknowledgment of the impending metropolitanization of the rural and regional countryside. Acknowledging the existence of a rural and regional culture, with its own value, coincided with the aims of the regionalists.[13] Lewis Mumford's words in the 1925 Regional Plan Number of the *Survey Graphic* resonated clearly. He wrote:

> Finally, regional planning is the New Conservation – the conservation of human values hand in hand with natural resources. Regional planning sees that the depopulated countryside and the congested city are intimately related; it sees that we waste vast quantities of time and energy by ignoring the potential resources of a region, that is, by forgetting all that lies between the terminal points and junctions of our great railroads. Permanent agriculture instead of land skinning, permanent forestry instead of timber mining, permanent human communities dedicated to life, liberty, and the pursuit of happiness, instead of camps and squatter-settlements, and to stable building, instead of the scantling and falsework of our "go-ahead" communities – all this is embodied in regional planning.[14]

3. THE SKYLINE DRIVE AND THE SHENANDOAH NATIONAL PARK

Planning the Skyline Drive involved the planning and opening of the Shenandoah National Park, so in many ways, the stories of the Drive and the Park overlap. The earliest references to the Drive and the Park come out of the Southern Appalachian National Park Commission report. In 1924, this Commission recommended the designation of the Shenandoah National Park, including, as mentioned earlier, the possibility of a "sky-line" drive.[15] Of note, the Southern Appalachian National Park Commission included Harlan P. Kelsey, a landscape architect, nurseryman, and former president of the Appalachian Mountain Club from Massachusetts.[16] Kelsey collaborated with the National Park Service throughout the early history of the Shenandoah National Park. From an early stage, Kelsey opposed the introduction of roads and "skyline" drives into national parks. Over time, however, he worked with the Park Service on road design and served an important role in placating the concerns of anti-road activists.[17]

From 1925 onward, a series of Congressional actions led to the designation of the Shenandoah National Park, and the Commonwealth of Virginia assumed the responsibility for securing the deeds to the land required for the Park. Governor of Virginia Harry Byrd created the State Commission on Conservation and Development in 1926 to obtain, among other things, the land necessary for the Shenandoah National Park.[18] The Commission on Conservation and Development, headed by a strong supporter of Byrd, William E. Carson, picked up where a park-supporting business organization, the Shenandoah National Park Association, had left off. Carson's Commission solicited and collected pledges for the Shenandoah National Park Association, made on behalf of the proposed

Shenandoah National Park. State legislation empowered this Commission to purchase property outright or through condemnation proceedings, using funds collected from the pledge drive.[19]

Between 1926 and 1929, the Virginia General Assembly and the Virginia courts granted the State Commission on Conservation and Development (SCCD) additional powers. The General Assembly passed legislation that allowed the SCCD to acquire land for the park through "blanket condemnation" proceedings, rather than one property at a time. This act led to eight condemnation proceedings – one action in each of the eight counties having land within the proposed park. By October 1929, the Virginia courts had upheld the constitutionality of this process. The legislation ultimately allowed the SCCD to acquire the necessary deeds to properties owned by mountain people, which would later be turned over to the National Park Service. In addition to the increased powers granted to Carson's Commission in 1928, the Virginia Assembly appropriated funds to purchase land from its owners. The direct appropriation of funding for the purchase of land is something that Byrd and the General Assembly had resisted in the earliest years of park development.[20]

By the end of 1929, the SCCD was still having difficulty raising the necessary money to purchase the land for the proposed park. Hope for additional private donations had dried up with the onset of the Depression. The original federal proposal for the park called for a minimum of nearly 385,500 acres. A 1927 report by the Assistant Director of the National Park Service, Arno B. Cammerer, called for a reduction in acreage to a minimum of 327,000 acres.[21] Congress ultimately amended its original park legislation to conform to this report.[22]

Interior Secretary Hubert Work, at the request of Virginia's Governor Byrd, had authorized the Cammerer report of December 21, 1927. Byrd was concerned that the Commission on Conservation and Development would not be able to raise the funds necessary to purchase the minimum land required by Congress in the original legislation authorizing the Shenandoah National Park.[23] In addition to lowering the minimum required acreage, the Cammerer report made recommendations on roads within the park. This report discussed roads for the first time since the Southern Appalachian National Park Commission recommendations of 1924. Cammerer noted that the proposed park, as originally envisioned, consisted of a "strip" of the Blue Ridge Mountains. Two main roads crossed the mountains in the designated park area – the Spotswood Trail and the Lee Highway.[24] Of note in Cammerer's report was discussion of the possibility of a "skyline" road within the boundaries of the park. Cammerer recommended that discussion of the road only commence once the land had been handed over to the Park

Service, and not before. His approach allowed for the possibility of survey, assessment, and then action.

Cammerer continued to promote the finest features associated with the proposed park. He wrote, "[I]n my opinion this park will be popular because of the unexcelled opportunities for camping on top of the ridge in the summertime when the lowlands are sweltering in the heat, and possibly to a large extent in this it will differ from all the other national parks."[25] The ridgeline made perhaps the greatest impression on Cammerer and he viewed it as a place for recreation – a place to camp and escape the summer heat.

3.1 The Hoover Camp and the Beginning of the Skyline Drive – William Carson Introduces President Hoover to Madison County

The Presidential Election of 1928 sent Herbert Hoover to the White House. Although he was a Republican and Virginia was a one-party Democratic state, Shenandoah National Park supporters received a boost from Hoover's election. Hoover had served as the president of the National Parks Association, whose December 25, 1924, *National Parks Bulletin* praised the Shenandoah National Park proposal and urged its members to support the project.[26] Moreover, Will Carson, who had met President Hoover over a decade earlier during World War I, knew of Hoover's involvement with the parks movement and his penchant for fishing and hiking. Soon after the election of 1928, Carson worked with President Hoover's Secretary, Lawrence Richey, and other Washington-based outdoorsmen who knew of the Blue Ridge possibilities, in an effort to pique Hoover's interest in the area. By the spring of 1929, Hoover had twice visited a possible "camp" site for a retreat, and Carson had taken time to secure, for Hoover's entertainment, fishing rights along the Rapidan River on the eastern slope of the Blue Ridge Mountains.[27] Clearly, Carson believed that the presence of the President of the United States within the boundaries of the proposed Shenandoah National Park would only raise public awareness and positive publicity for the park movement. Carson duly went to great lengths to smooth the way for the construction of Hoover's camp.[28]

3.1.1 Discussion of Other Roads

In 1929, President Herbert Hoover established a vacation "White House" along the Rapidan River, later known as the Hoover Camp, within the proposed park boundaries. The move aided the floundering park effort.[29]

Not only did the Hoover Camp bring notoriety to the proposed park, it also brought roads to the area in 1929.[30] The talk of roads along the crestline had been revived for the first time since the Southern Appalachian Park Commission report of the early 1920s.

The President began using the camp on the Rapidan River in the autumn of 1929. Roads connecting the Hoover Camp to the small Madison County town of Criglersville to the east were completed by the Virginia State Highway Department soon after. At this point, seeing the value of additional roads within the proposed park and the publicity it would bring, Carson asked the State Highway Department to estimate the cost of a road from the President's camp up the Rapidan River valley to the ridgeline, then northward to the mountain camp known as Skyland (owned by a Park supporter named George Freeman Pollack), and then north along the ridgeline to the Lee Highway at Thornton Gap.[31] The Virginia State Highway Department never built the road to Skyland as proposed by Carson. Undaunted, Carson worked to first get authorization for a road through the proposed park and then worked to secure federal funding for the other segments of the road that would connect Hoover Camp to Skyland, believing the road construction could help move along the Shenandoah National Park project.[32]

3.1.2 Federal Appropriation for Roads

The Depression and a lingering drought in Virginia had caused severe economic difficulties in Central Virginia and the Shenandoah Valley. Shenandoah National Park supporters and the Virginia Congressional delegation obtained emergency relief funds to remedy the economic conditions afflicting the citizenry. As chairman of the Virginia Commission on Conservation and Development, Will Carson not surprisingly viewed economic development as a main priority of his work. In an effort to promote the Park idea and provide economic opportunity for the people affected by the economic downturn, Carson set about to revive the "skyline drive" idea recommended in the 1924 Southern Appalachian National Park Commission report. He also continued to argue for the connection between the President's camp and the Skyland Resort on the crest of the Blue Ridge. The sagging economy and the prospect of relief funds allowed Carson to change the focus of his appeal for roads. Instead of arguing that the road would connect the President's camp to Skyland or that construction of the road along the ridge would carry out the idea called for in 1924, he emphasized that road construction would aid the local economy and "probably mean the saving of many lives from starvation."[33]

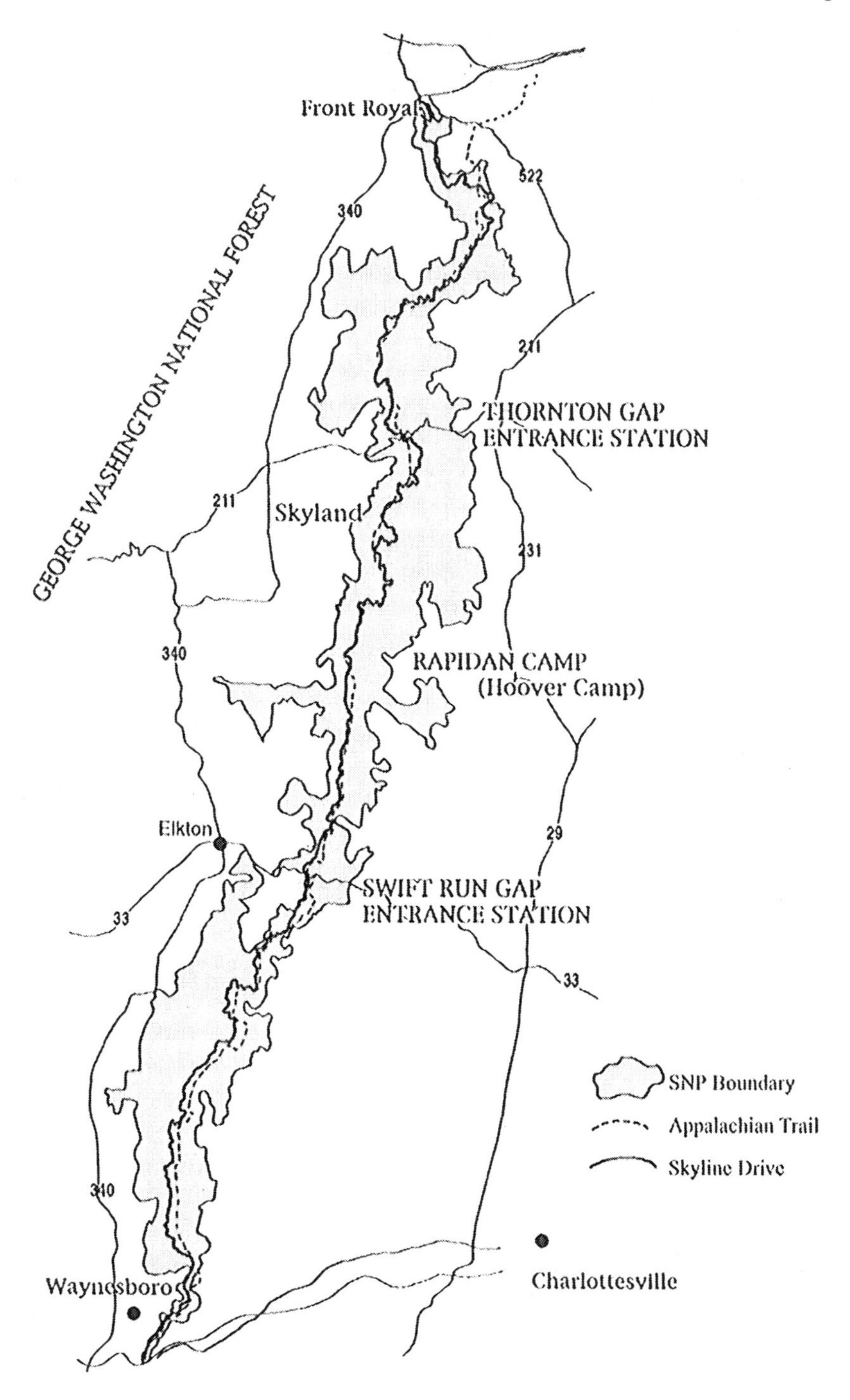

Figure 5.1. The Shenandoah National Park, Skyline Drive, and the Appalachian Trail.

As a matter of design and topography, the relief appropriation was to fund a road from Hoover Camp, up to the ridgeline, then northward to an improved highway crossing the Blue Ridge. Specifically, it was to run from the President's camp nearly perpendicular to the north-south run of the Blue Ridge Mountains. That is, the proposed park had a linear orientation (north and south) and the road from the President's camp along the Rapidan to Skyland would have provided access up the east side of the mountains. Once on top of the mountain, the road would run northward along the spine of the Blue Ridge. It was then set to progress northerly to Skyland and then to Panorama and the Lee Highway at Thornton Gap.[34] At Thornton Gap, the ridgeline road would intersect with the Lee Highway, an improved road that crossed the Blue Ridge at a gap in the mountains and connected directly to Washington. This proposed route provided, according to Carson, a quicker route for the President's return to Washington, DC.[35] (See Figure 5.1 for the basic orientation).

Initially, Carson worked with Virginia Senator Carter Glass to get additional funds for the roads through the "Roads and Trails" section of the National Park Service appropriation (under the Department of the Interior) in the Emergency Employment Bill of 1930. Horace M. Albright, Director of the National Park Service, also worked for this appropriation, despite the fact that the land for this road had not yet been condemned and turned over to the Park Service.[36]

Ultimately, Carson, Albright, and the Virginia delegation secured funding for the road project through the Emergency Public Works Act of 1931. According to the Department of the Interior "Memorandum for the Press" of February 4, 1931, Secretary of the Interior Ray Lyman Wilbur allocated to the Shenandoah National Park road project $250,000 of the $1,500,000 total Park Service appropriation.[37] Leading up to this decision, Carson, Governor John Garland Pollard (the new governor of Virginia), and the members of the Virginia delegation had worked with President Hoover, the Secretary of the Interior, and the President's personal Secretary to secure the funding.

The argument for the funding was broad and wide-ranging. On December 31, 1930, anticipating the possible availability of funds for Shenandoah through the Emergency Employment Act, Carson wrote to Representative Louis Cramton of Michigan:

> The Conservation and Development Commission has been pressing the job of acquirement of the lands in this section [the proposed Shenandoah National Park area], and we hope by the end of this coming year the work will be completed.[38]

Carson felt that the State of Virginia had no power to address the economic difficulties caused by the Depression and the drought. In the same letter, Carson emphasized the importance of the road project in the development of the Shenandoah National Park. He continued to Cramton:

> What I want to do is interest you in inserting in the first Deficiency Bill a clause that would allow the building of a road from Panorama on the Lee Highway via Skyland to an intersecting point in the Rapidan Valley road, out of drought relief funds allocated to the National Park Department. This road would develop the most scenic section of the Shenandoah National Park, is about midway of the Park and is entirely within the proposed Park area [. . .].[39]

Carson's plea indicated a couple of things. First, the road project would ultimately be incorporated within the National Park, despite the fact that, according to the Arno Cammerer Report of 1927, no plan for roads had yet been established. This was a stretch and intended to solidify his argument. Surely, Cramton did not know of the Cammerer report. Second, Carson implied that the road between the President's camp and the ridgeline at Skyland had been built as of December 1930. This was not the case. The road between the President's camp and the ridgeline near Skyland did not get surveyed until the spring of 1931.[40] Further, by the end of the spring of 1931, the Bureau of Public Roads and the National Park Service abandoned the idea of constructing a high quality road between the President's camp and the ridgeline near Skyland.[41]

Nearly all of the Virginia delegation to Congress, and the Governor of Virginia, pushed for the allocation of these funds. Carson called upon local citizens involved in the Park movement to make their pleas well known. And they responded: a local physician, Roy Lyman Sexton; a realtor and lumberman, Ferdinand Zerkel; and the owner of the Skyland Resort, George Freeman Pollack, all wrote testimonies to the dire straits of the people in the region.[42] Ironically, these park boosters pushed for the development of a project that would do more damage to a way of life than an economic downturn or a periodic drought. Moreover, many of the mountain people who did get jobs through the drought relief appropriation worked on a road that would ultimately lead to their own displacement.

The federal drought relief funds used for this road construction were at times referred to as appropriations from the Emergency Public Works Act of 1931 and as the First Deficiency Act of 1931. The appropriated funds were assigned to the Bureau of Public Roads through the National Park Service and the Interior Department. Once the deeds to the land within the proposed right-of-way were conveyed to the National Park Service, the Bureau of Public Roads intended to begin constructing the road. Under an interagency agreement, the National Park Service provided technical advice to the Bureau of Public Roads whenever proposed roads were set for construction in a national park.[43] Unfortunately, since the assignment of the right of way for the road did not match the minimum number of acres required for the establishment of the park, the Park Service and the Secretary of the Interior ran into difficulty when the expenditure of such a large amount of money was made public nationwide. In other words, the Park Service owned the land over which the road would be built, but they did not yet have the minimum acreage for the park.

3.1.3 McKellar's Concerns

A May 3, 1931, "Letter to the Editor" printed in the *Knoxville News-Sentinel* decried the fact that Virginia reaped the benefit of the "relief" funds, while the citizens of Tennessee did not – even though the economic conditions in the eastern part of Tennessee mirrored those in Virginia. Further, the writer questioned the outlay of funds within a proposed park area even though the park had not yet been established. Given the similarities in the economic conditions, the writer questioned his local Congressman's ability to secure comparable funding for his home state.[44]

The editorial response to this letter drew both of Tennessee's senators into the debate, although the editor stated that the true "proper policy for the local" Congressional delegation would be to "support the position of the National Park Service with respect to the park."[45] Needless to say, the letter caused Senator Kenneth McKellar and Congressman J. Will Taylor of Tennessee to write to the Secretary of the Interior and the Director of the National Park Service for clarification. The responses from Director Horace M. Albright and Assistant Director Arno B. Cammerer spanned the spectrum of political rhetoric. The rhetoric, however, reflected the direction the National Park Service believed the road-building project would take the Shenandoah National Park and ultimately, the Skyline Drive.

In a personal letter to the editor of the Knoxville *News-Sentinel*, National Park Service Director Albright praised the editorial response to the initial letter. Noting the stalled nature of the Shenandoah National Park, Albright said of the road project, "[i]t will help put over the project if people can get up on the top and see what there is to the area."[46] To Tennessee Congressman Taylor, Secretary of the Interior Ray Lyman Wilbur argued that the Congressional relief appropriation required that the Bureau of Public Roads spend the funds by June 30, 1931. Since the road could only follow one route – the ridgeline within the proposed Shenandoah National Park – the Park Service decided to spend the money there. The planning required to map out a route within the Great Smokies project would preclude the expenditure of these relief funds at that proposed park in the time allowed. Besides, he wrote,

> There are already numerous roads for motor traffic into the Great Smoky Mountains National Park area, and because of the very nature of the topography new road development of any kind by the Federal Government can only be undertaken there after the whole park has been thoroughly studied for the best locations by the landscape architects and civil engineers of the National Park Service. On the other hand, the Shenandoah area to be traversed is at present entirely inaccessible by motor vehicle; there is only one practicable route for any road to take along the narrow skyline.[47]

Wilbur and Cammerer – who sent the same exact letter to Senator McKellar on the same date[48] – essentially claimed that the initial road building project within the proposed boundaries of the Shenandoah National Park involved no planning or engineering. Further, Wilbur and Cammerer used language implying that the Shenandoah National Park area was inaccessible from the outside – a point which was untrue since two primary roads bisected the area at Thornton Gap and Swift Run Gap. Their point was based on a technicality – a road from the President's camp on the Rapidan River into the proposed park area did not yet exist.

McKellar was still not satisfied with the expenditure of funds for a park that had not yet been turned over to the National Park Service. Of the Cammerer letter, he wrote with no small amount of sarcasm:

> I judge from your letter, however, that the projectors of the Shenandoah Park must have met the Government requirements or certainly there would not be funds expended for roads in a park that was not yet a park.[49]

In an incredibly bureaucratic response, Cammerer replied:

> It was not my intention to lead you to believe that the projectors of the Shenandoah Park have met Government requirements as to park creation. None of the three proposed eastern national parks projects has yet done that. I desired to bring out in my letter that none of these projects has as yet secured national status and that no road has been projected in any one of them as a park road or from park funds.[50]

Yet the road-building on the ridgeline within the proposed Shenandoah National Park continued throughout the spring and summer of 1931.

Finally, the Editor's response to the initial letter criticizing Congressman Taylor must have hit home with Albright, Cammerer, and even Secretary Lyman. The Editor had written:

> Let us not be envious of the Virginians because they have obtained a road. The Great Smokies are so outstanding, so unique, that roads cannot be run thru [sic] them hastily without spoiling them. The roads must be carefully planned, so that they will run in such places and in such manner as will not spoil their wild grandeur. A single ill-advised road could spoil the park irreparably. What nature has taken millions of years to build must not be damaged by man in a moment.[51]

The National Park Service, in conjunction with the Bureau of Public Roads, had discussed the design options available during the late winter of 1931. Unfortunately, the planning and design consultation was haphazard, rushed, and subsumed by two of the arguments made by Wilbur, Albright, and Cammerer: the funds had to be spent quickly, and the road was built in part to raise excitement for the project.

3.1.4 Thomas Vint, Gilmore Clarke and the Planning of the Skyline Drive

The "relief" funds allocated for the road-building provided a unique opportunity for park proponents to construct the "sky-line" drive proposed earlier. Unfortunately, because the availability of funds had a deadline, the road/parkway designers with the National Park Service and the Bureau of Public Roads had very little time to reorient their plans, survey the area, and begin construction. Similarly, the public's ability to comment on the construction work had passed, as had the opportunity for comment by park proponents concerned about the impact of a "sky-line" drive.[52]

And finally, the "relief" funds and their appropriation made the earliest planning of the Skyline Drive seem as though it was anything but planning. It is possible to make the argument that the idea of constructing a "sky-line" road was low on the list of reason to build roads in the Shenandoah area and that the initial appropriations rivaled subsequent New Deal work programs. That is, the most compelling reason to begin the construction of the Skyline Drive was the employment possibilities. Again, quoting the Department of the Interior press release of February 1931, "The main purpose of building the highway at this time, however, is to give the mountain people in this area an opportunity to work and provide food and clothing for themselves and their families."[53] A later press release read, "The people of this mountainous area are in dire straits [. . .]. Like all mountain people they have a sturdy independence. The construction [. . .] at this time will [. . .] offer them employment instead of charity."[54]

In March of 1931, Thomas Vint, Landscape Architect for the National Park Service, traveled to the East Coast to consult on the Shenandoah project, as well as the parkway projects in Tidewater (Colonial National Parkway) and the Washington, DC, metropolitan area (the Mount Vernon Parkway). At this point, although roadwork along the ridgeline of the Blue Ridge had been funded and at least talked about, the Bureau of Public Roads had only commenced its engineering survey of the route.[55] Will Carson had asked some of the local park supporters to aid in scouting a preliminary line along the ridge. Self-proclaimed "park-nut" Ferdinand Zerkel had obliged and sent his proposed route to Carson. He concluded his letter by referring to the Chief of Construction at the Bureau of Public Roads, H.K. Bishop. In an indication that as of the end of January 1931, the Bureau of Public Roads (not the Park Service) remained in complete control of the surveying and siting activity, Zerkel wrote of his proposed route: "Of course, Mr. Bishop will likely scale this more closely on the maps even ahead of any field check up of route, etc." [56]

In light of the then-recently approved appropriation for the road along the ridgeline, the Park Service had finally begun to look into some sort of plan for the route. Interestingly, although funds had been approved, Cammerer still wrote of the road as "not yet a settled thing," and worried about too much publicity.[57] Indeed, the Potomac Appalachian Trail Club had begun to indicate some displeasure with the proposed road along the ridgeline. Zerkel wrote to Cammerer:

> One other thing, while chatting with you on my typewriter, I have heard that certain officials of the Potomac Appn. Trail Club thought the Skyline Road something to disturb their privacy and otherwise injure the Park from their own viewpoint. Now, Friend "Cam.", I am a member of the Club and have some fine friends in it whom I would hesitate to lose.[58]

In the end, however, Park Service officials finally made it to the ridgeline to inspect the proposed route of the road for which they had received funding.

In mid-February 1931, Park Service Director Horace Albright, Assistant Director of the Park Service A.E. Demaray, as well as a number of local Park supporters and PATC members, toured the ridgeline from Skyland to Thornton Gap – perhaps one of the earliest tours by the Park Service officials. After inspecting the Shenandoah area with the Assistant Landscape Architect for the Park Service, Charles E. Peterson, Vint wrote a memo to the Director of the National Park Service, Horace Albright, recommending a cessation of the plans for a high-standard road to link the President's camp with Skyland. He stated that the ridgeline road between Swift Run Gap and Thornton Gap should receive the fullest amount of time, effort, and money possible, while the existing lateral roads through the gaps were deemed sufficient. In a memo that seemed to answer many of the questions regarding the road effort, Vint wrote:

> 1. It is essential that an agreement be reached as to the general plan of the main road system of the Park. With this in mind, all work that is done on roads will be toward one solution.
>
> 2. My study of the plan to date indicates that the park road system will center around a main tourist road along the backbone of the Blue Ridge Mountains.[59]

The Vint memo settled a number of things. First, it roughed out a plan – although the Knoxville *News-Sentinel* Editor's criticism noted above resonated six weeks later. Second, it outlined a master plan for roads within a national park not yet established. Third, it settled the lateral road concern and turned all effort towards the Skyline Drive. Fourth, it explained the difference between work on park roads in the West and parkways in the national parks in the East. Parkways in the national parks in the East would serve as recreational outlets in and of themselves. Finally, Vint paid homage to the Park Service's duty to "preserve and protect the natural and wilderness conditions" within a park's boundaries.[60] He recognized the intrusiveness of the road.

The Park Service, the Bureau of Public Roads, and President Hoover, personally, all agreed with Vint's recommendations. A press release issued by the Department of the Interior on March 25, 1931 stated:

> One of the most scenic highways of the country that may ultimately extend for 150 miles, is to be built along the summit of the Blue Ridge Mountains throughout the length of the Shenandoah National Park area in Virginia [. . .]. The proposed highway eventually will follow the skyline of the ridge from Front Royal to Rock Fish Gap [. . .].[61]

To carry out the plan, Vint and Albright called upon Charles E. Peterson, Assistant Landscape Architect in the Park Service. Peterson, already working in the Tidewater area on the Colonial Parkway project and other East Coast projects, came to spend about one-third of his time working on the Skyline Drive project.[62]

3.1.5 Gilmore Clarke Gets Involved

On March 28, 1931, A.E. Demaray sent H.K. Bishop of the Bureau of Public roads six copies of the plans of the route for the Skyline Drive through the proposed Shenandoah National Park. Bishop responded in a way that spoke to the difficulty and, indeed, the newness associated with the Skyline Drive. Bishop wrote, "Why didn't you call it a boulevard – that's what it is going to be."[63] In other words, the road would not mirror other park roads built by the Bureau of Public Roads under the direction of the Park Service, as had been the case out West. Further, by using the term "boulevard," Bishop implied that the Skyline Drive would take on a more intrusive role than most park roads out West.

In a series of events that proved even more informative of the nature of the Skyline Drive, Gilmore Clarke, Landscape Architect with the Westchester Parks Commission in New York and formerly employed by the Bronx Parkway Commission, commented on the early design work for the Skyline Drive. While in Washington, DC, Clarke had the opportunity to view the initial plans for the Skyline Drive, but, as of the date of his letter – July 6, 1931 – he had not yet had the opportunity to view the site itself. Of the plans he viewed, Clarke declared plans "showed that no attention whatever had been given to refinements of alignment which should obtain on a road or drive in our National Parks."[64]

Clarke's concerns were obviously design-oriented and, most likely, valid. By the 1920s, the Park Service, in partnership with the Bureau of Public Roads, had apparently abandoned the curve and tangent method of design and construction on park roads. Instead, landscape architects and the engineers with the Bureau of Public Roads had started employing "curvilinear stretches interconnected with radial curves."[65] Basically, this new method allowed the driver to ease into curves slowly, rather than linking straight lines with sections of circles. This new method of dealing with curves had not yet been institutionalized throughout the Bureau of Public Roads. Having received word of the Clarke visit in Washington, DC, Assistant Landscape Architect Peterson penned a letter to Horace M. Albright. He wrote:

> [An official with the Bureau of Public Roads] had shown C & D [Clarke and Jay Downer – also on the trip] the plans for the 42-mile Shenandoah jobs that Bishop's outfit has just done and it seems that they didn't like them worth a damn. They think that tangents should spiral into curves and *vice versa*. This is a method different from what the BPR [Bureau of Public Roads] does for us in the west, (and quite possibly is better from the landscape standpoint).[66]

Clarke's involvement incited hastily sent letters of veiled accusation and defensiveness between Peterson, Albright, Demaray, Clarke, and Thomas MacDonald of the Bureau of Public Roads. Although Clarke had significant stature within landscape architecture professional circles and within the National Park Service itself, he also had connections with the Rockefeller family, and that position had to take precedence. In establishing Acadia National Park and Great Smoky Mountains National Park, the National Park Service had received generous funding from the Rockefellers.[67]

Once Peterson had received a copy of Clarke's initial letter addressed to A.E. Demaray in Washington, DC, Peterson defended his own work and that of the Landscape Division of the Park Service. (Prior to receiving a copy of the Clarke letter, he had only heard of the criticism through engineers with the Bureau of Public Roads. At that point he was not aware that Clarke had directed his criticisms at the Landscape Division.) Peterson began his response somewhat sarcastically noting that, of course, the plans look different from the Westchester parkways – which he characterized as "suburban."[68] Peterson made the point that the design of the parkway through Shenandoah would, by necessity, be different from the suburban parkways of Westchester County. The method for producing the design of the Skyline Drive would, therefore, be different, Peterson continued:

> While it would appear in the letter [Clarke's letter] that the landscape architects had failed to fulfill their normal function of insisting on a proper design, I am persuaded that the situation cannot be so simply and easily outlined [. . .]. In light of the mountain road survey which I have been concerned with and using the location policies of the Bureau of Public Roads as practiced in the West as a basic premise, the line I inspected seemed to me to be well laid out; and I so reported in a letter to Mr. Bishop. If this road is not laid out, then we have no good roads in the western Parks. I would hesitate to make such an indictment, because I believe that the Landscape Division of the Park Service has had much more experience on this mountain type of work than all of the other landscape offices which have ever existed in this country.[69]

Peterson continued by noting that the lack of landscape architectural help impeded the ability of the Landscape Division to complete such a thorough pre-construction regimen.[70] Peterson noted the need for further engineering on the Drive design. His call for additional topographic detailing created more work for the Bureau of Public Roads engineers, but certainly added to the information available for planning the road. Moreover, the in-depth study and design methodology called for by Peterson was in effect an admission that the earliest design and plans for the Drive were done without all the necessary data available.[71] That, in a way, reinforced the concern that had been voiced by the Editor of the Knoxville *News-Sentinel* earlier that spring.

While Peterson's comments indicated that Gilmore Clarke's concerns reflected disputes within the context of design, a broader view demonstrates that the conflict actually existed on a number of levels. Clarke tried to bring to bear the sensibilities of a suburban parkway designer. Peterson reacted and showed that the two were different – professionally, institutionally, and geographically. For all intents and purposes, Gilmore Clarke was a metropolitan planner of parkways. His work and career established this. Admittedly, however, Peterson did not fall neatly into the "regionalists" camp.

During the late 1920s, Clarke collaborated with the Park Service on work at Mammoth Hot Springs and, in later years, on the Mount Vernon Memorial Parkway. In addition, members of the Landscape Architect's office of the Park Service worked with Clarke in Westchester County during 1930 and 1931.[72] And Park Service officials concerned with the road construction project in the Shenandoah had respect for Clarke and tended to take what he said seriously. Clarke ultimately wrote back in September of 1931 expressing his confidence in the ability of the Park Service and the Bureau of Public Roads to remedy his concerns.[73]

Linda Flint McClelland, in her book *Presenting Nature: The Historic Landscape Design of the National Park Service, 1916-1942*, places much emphasis on Gilmore Clarke's influence with the Park Service during this time. In addition to arguing that parkways in national parks were recreation opportunities in and of themselves (motoring as recreation), McClelland notes Clarke's influence by showing that these parkways combined ideas employed in western parks with those of the suburban parkways of Westchester County.[74]

Clarke's concerns and Peterson's responses during the planning phase of the Drive illustrate how the design and construction of the Skyline Drive broke new ground within the profession. It also demonstrates that the history of this project does not effortlessly fit into the previous histories of parkways (the historical progression discussed earlier).[75] One of the larger issues brought about by the exchange between Clarke and the Park Service over the Skyline Drive was the very nature and purpose of parkways and park roads within the Park Service's mission.

3.2 Transition

With the design and engineering phase under way, Will Carson and the State Commission on Conservation and Development had turned to the condemnation of rights-of-way for the road itself. Carson sought to speed up the acquisition of the land for the Drive – at least the parcels necessary for road construction – prior to acquiring the land necessary for the park.[76] Carson accomplished this in time for the road construction to proceed in a timely manner.

3.2.1 Community Displacement

The designation of the initial road project as a "drought relief" measure let the SCCD and the Park Service circumvent the requirement by Congress that the deeds be secured before funds could be spent on a national park. The acquisition of the deeds needed for the first section of the Drive went smoothly because the SCCD negotiated the purchases,[77] although there were some instances that caused concern.[78]

For the SCCD, the concern over land condemnation amounted to two primary issues: could the Commission raise the money needed to pay for the condemned land, and could the court proceedings be negotiated in a manner favorable to the Commission. Outside of land owned by well-known families in Virginia, landowners for the most part were treated as inconsequential nuisances. In a response to a Charles E. Peterson memo about connecting the southernmost point of the Skyline Drive to a main road near the southern boundary of the Shenandoah National Park, Arno B. Cammerer wrote, "[thc area] is covered with estates such as that of Tom Scott of Richmond that precludes going straight through to the Rock Fish Gap."[79]

Well-off Richmonders such as "Tom Scott" did not seem to suffer from the negative consequences of blanket condemnation proceedings. It was the group of people designated as "mountain people [with] a sturdy independence" that suffered most from the Skyline Drive's development. Further, the impact of the construction of the Drive was additional land condemnations for the establishment of the Park.

The proponents of the Shenandoah National Park concept for the most part saw the beginning of the construction of the Skyline Drive as a catalyst for the Park. The acquisition of public lands for recreation, conservation, and preservation often carried with it a positive connotation; this was not the case in central Virginia. The idea of preserving the primeval wilderness along the Appalachian Mountain chain appealed to and inspired Benton MacKaye, for instance, but his idea of social reform and regional planning did not include displacement of indigenous communities. Some of Benton MacKaye's early writing called for federal intervention, yet MacKaye never wrote of displacing communities to accomplish the aims of his regional planning initiative.[80]

3.2.2 Appalachian Trail, the Potomac Appalachian Trail Club and the Drive

As work on the Drive continued on the first section from Skyland to Thornton Gap, disputes over the relationship of the road to the Appalachian

Trail and the ridgeline surfaced. The Potomac Appalachian Trail Club (PATC) had formed in 1927, in part to promote the Shenandoah National Park idea, and in part to continue the Appalachian Trail through Virginia. The actions of the PATC were as important for what it did do as for what it did not do. Many PATC members were Park supporters first and Appalachian Trail supporters second. On the other hand, the strong AT supporters viewed the Skyline Drive in various ways. The "true believers" consistently condemned the Drive. "The Skyline Drive, because of the nature of the terrain, does not of course literally follow the route of the trail, but it intersected the trail at so many points that what resulted was its practical obliteration."[81] The pragmatists within the PATC, on the other hand, accepted early on in the process of constructing the Drive the decision to build the Drive, despite the lack of planning and opportunity for comment. Two early PATC members who also accepted the Drive for pragmatic reasons commented:

> The construction of the Skyline Drive in the park area was authorized practically without notice; no opportunity was afforded for discussion of the general advisability of the undertaking.[82]

While this is a major point, those members of the PATC who accepted the Skyline Drive pushed the positive benefits – easier access for Trail maintenance and the aid of the Park Service and the Civilian Conservation Corps (CCC) when the Trail had to be moved from its original line to accommodate the Drive.[83]

3.3 Conclusion – History

In sum, the earliest planning of the Skyline Drive began with political maneuvering by the chairman of the State Commission on Conservation and Development, Will Carson, with the willingness of the federal government to participate in a work program. Carson united politicians on all levels of government, rallying around a project that would be characterized as a "work project."[84] In the end, however, the road-building project led to the Skyline Drive and served as a catalyst for the establishment of Shenandoah National Park itself. As the SCCD chairman, Carson headed a park movement that was in serious jeopardy by 1929. The road-building project appeased critics and heartened proponents alike. Further, as discussed below, the business groups behind the establishment of the park and ultimately the Skyline Drive viewed roadways as a tourist attraction and a way to bring money to Virginia. The Skyline Drive actually accomplished both. A brief history of the Skyline Drive written in 1935 under the auspices of the SCCD read:

> The conservation chairman [Carson] argued that it was impossible to sell the Shenandoah National Park to the nation without showing it and that it was difficult to show the park without a roadway through it.[85]

The drought-relief funds and the prospect of a "sky-line" road breathed new life into the Park project and into the region, which were both at a crossroads. The condemnation proceedings by the Commonwealth of Virginia had been held up; Congress had said federal funds could not be used within the boundaries of the proposed park until deeds to the land had been turned over to the Park Service; the Depression had begun to catch up with the economy in Virginia, including the Shenandoah area; and the region was experiencing a severe drought.

The central section of the Drive officially opened in 1934 (although visitors were allowed to drive the first section periodically as early as 1932), followed by the northern section in 1936, and the southern section in 1939. At that point, the Drive measured 97 miles and was upped to 105 in 1961 when a portion of the Blue Ridge Parkway north of U.S. Route 250 near Waynesboro was added to the Skyline Drive.[86]

4. THE CONFLICT

The history of the early construction of the Skyline Drive did not adequately reveal the underlying battle between those who believed in the regional vision and the parkway boosters who believed in facilitating economic development. The earliest members of the Potomac Appalachian Trail Club were strong Shenandoah National Park supporters, while early Park and Skyline Drive supporters also belonged to the Potomac Appalachian Trail Club.[87] That is, the membership of the PATC overlapped with the Park and Drive booster groups. Before the start of construction of the Drive, the talk of roads in the proposed Shenandoah National Park had been minimal (after the 1924 Southern Appalachian National Park Commission had made its recommendation). When the Bureau of Public Roads and the National Park Service began construction of the Drive in 1931, some PATC members followed their economic development interests, some merely accepted the Drive and worked to reroute the AT, while others consistently worked to privilege the AT over the Skyline Drive.

Benton MacKaye's proposal for his personal involvement in the early park movement foretold the ensuing conflict. In 1925, Benton MacKaye wrote Harlan P. Kelsey, one of the members of the Southern Appalachian National Park Commission, to offer his services to the Commission as both surveyor of the recreational possibilities in the proposed Shenandoah

National Park and as a surveyor of that section of the Appalachian Trail. In an interesting exchange, Kelsey first told MacKaye that he would take up his offer with the rest of the Commission. Kelsey further noted that the development of the Appalachian Trail and other recreational developments "should come after the Park is secured."[88]

Before engineers with the Bureau of Public Roads laid the stakes marking the line of the road in the spring of 1931, both the regionalists and the economic boosters were hard at work promoting their own ideas for the region. The Potomac Appalachian Trail Club continued to build the Trail and shelters, creating a recreational outlet accessible to East Coast urban centers, as Benton MacKaye had envisioned. The boosters, meanwhile, talked of roads and the likelihood of an economic boom should the Shenandoah National Park include roads that would ferry tourists and dollars up and down the Shenandoah Valley and central Virginia.

4.1 The Regionalists – Motivations and Actions

The regionalists' vision for the southern Appalachians and the prospect of a Skyline Drive began with Benton MacKaye's 1921 article, "An Appalachian Trail: A Project in Regional Planning." Of course, between 1921 and the founding of the Potomac Appalachian Trail Club in 1927, the construction of the Appalachian Trail did not follow MacKaye's vision as completely as he may have liked.

The Appalachian Trail did, however, at a minimum provide an alternative to the metropolitanization of the cities and suburbs of the eastern seaboard. MacKaye recognized early on that his vision might not be implemented exactly as he planned. The Appalachian Trail and its supporters, from their founding through the battles with the Park Service and road boosters, believed that the Trail would provide an alternative to the metropolitan environment. The conflict with the road boosters not only included the issue of the Skyline Drive "obliterating" the Trail, but also the lengths to which Trail supporters would go to prevent the construction of the Skyline Drive.

In 1925, Harlan Kelsey and Benton MacKaye corresponded about the newly proposed park, possible skyline roads, and the Appalachian Trail. Kelsey wrote to MacKaye:

> I am particularly interested in having the trail go along the crest of the Mountains in the Shenandoah National Park area rather than a skyline road which I believe would seriously detract from the ecological value of the Park. More than this I believe that the Park for the most part should be kept entirely wild and restored to its natural conditions as fast as

> possible. This means, of course, that the camp sites will be grouped rather than scattered over the entire area.[89]

Kelsey continued in a postscript:

> Of course this is matter of development and administration, but so many hold a different view it doesn't hurt to discuss the matter early in the game.[90]

By this point, the Southern Appalachian National Park Commission had made its initial report public and had indicated that a "possible sky-line drive along the mountain top" would be the "greatest single feature."[91]

MacKaye responded:

> I agree with what you say about maintaining the "ecological value" of this area. Your Commission seems to have in its hands the power to determine for this country what the word "Park" is going to mean – whether the preservation of a bit of the original North America or the making of one more Coney Island [. . .]. Are Parks going to be offsets to our civilization or adjuncts of it? Unless the first is deliberately worked for the other will inevitably occur. And the choice will have to be made, as you say, early in the game.[92]

The words that MacKaye chose here outline the conflict between the Skyline Drive and the Appalachian Trail. MacKaye viewed the Trail as an "offset" to an ever-industrializing civilization/metropolitanization and the Drive merely as an "adjunct" or a facilitator of metropolitanization.

Harlan Kelsey continually expressed reservations about even the earliest prospects of the Skyline Drive. Just after his exchange with MacKaye, Kelsey wrote of the narrowness of the proposed park and the damage a bisecting road would cause:

> With the Shenandoah National Park I know that some of my associates want a skyline road. That's good propaganda stuff and it would make a magnificent drive but the Park itself averages about 10 to 12 miles wide and it would split it right in two in the middle. There are two main highways crossing the Park in low gaps and I believe that from these gaps roads could go up to some of those Mountain peaks and that is about all for roads in the Park.[93]

Again, Kelsey not only disapproved of the road along the ridgeline proposed by some of his fellow commissioners, he truly believed in the preservation of "natural state" lands within national parks. Further, Kelsey felt that road development should be left to the administration of the Park once the land was set aside and deeded to the National Park Service.

Throughout the history of the construction of the Drive, the Park Service remained cognizant of Kelsey's views. In April 1931, Horace Albright responded to Kelsey's concern over the early construction of the Skyline Drive. Of course, even in the spring of 1931, the proposed park land had not yet been deeded to the Park Service, yet road development had begun. Albright suggested that the road would likely not go beyond Skyland (the initial construction phase) and might perhaps, depending on funding, "loop" down to the President's camp along the Rapidan.[94] Albright's comments came across as disingenuous. Thomas Vint had already proposed that the Park Service only pursue funding for the Skyline Drive, rather than any other lateral or circular roads. Albright no doubt knew that there existed a certain amount of dissent over the construction of roads along the ridge, and likely was attempting to soften the ultimate conflict. The Shenandoah National Park project was precariously balanced at the time. Although Kelsey was not in Washington then, his opinion on the roads resonated with members of the Potomac Appalachian Trail Club and the Appalachian Trail Conference. The PATC membership included many Washington, DC, professionals. Albright's language may have been crafted in deference to these members and their respective connections throughout the government.

By the spring of 1932, Benton MacKaye had, along with Kelsey, accepted the construction of the Drive between Thornton Gap and Swift Run Gap (the middle section of the Park). But MacKaye did not acquiesce to the construction in the northern section. In a letter to Arno B. Cammerer, MacKaye discussed the "future extension" of the "Skyline Drive" and acknowledged that Cammerer had known nothing about it.[95] MacKaye continued by noting that Harold Anderson, Secretary of the Potomac Appalachian Trail Club, had just recently seen a series of stakes marking a proposed road northward from Thornton Gap. MacKaye was suspicious of the stakes' purpose. He wrote, "Does this mean the 'Future Extension' that we feared? Does it mean the extension of the policy of motor skyline vs. foot-path skyline in the National Parks?"[96] Cammerer responded to Harold Anderson (MacKaye had asked him to do so) that the stakes only indicated a survey and did not necessarily mean that the Park Service would construct the road.[97]

As it turned out, Cammerer's response mirrored Albright's earlier reply to Kelsey in its disingenuousness. By the autumn of 1932, the Drive northward from Thornton Gap had not only been staked, but the line had been revised with the help of Harlan Kelsey and apparently approved by Park Service Director Horace Albright. In a letter to Albright, Kelsey once again expressed concerns over roads within the Park, but admitted to the necessity of the extension when he noted the early and evident congestion in the first section. Kelsey did, however, again plead for the Park Service not to build a road in the southern section of the Park: "I'm inclined to think the pressure [to build the southern section] is going to be irresistible."[98]

In a further discussion of this idea, MacKaye wrote an analysis of development along the Appalachian region. Picking up on themes central to his concept of the region, the primeval, communal, and rural environment and the automobile, MacKaye believed the Appalachian Trail – "the foot-path movement," as he called it in this unpublished analysis – through the primeval was the only way to stem the tide of unchecked metropolitan growth. While the Appalachian Trail protected the primeval, the highway – specifically the townless highway – provided for commerce, community, and the movement of people and goods. In the analysis laid out in his paper entitled "The 'Open Door' in America: Manifest Destiny for Appalachia, Not Manchuria," MacKaye argued that a skyline road would "pierce with urban influence" the outdoor recreation opportunities in the Appalachians and "kill" the Appalachian Trail.[99]

This "Open Door" analysis linked together most of the concepts MacKaye debated over the course of his professional life. Further, it clearly meshed his conception of regional planning with his opposition to the Skyline Drive. MacKaye refused to envision or comprehend the use of a roadway as a recreational option, at least at this point in his career. Roads were used for transportation of people and goods, hence the townless highway. The ridgeline and the wilderness of the Appalachian mountain chain had to be reserved for the rural (where appropriate development of agriculture existed) and the primeval (wild mountains and forests). This latter area, according to MacKaye, was to be accessible by the Appalachian Trail alone.[100] At the Appalachian Trail Conference held at Skyland in 1930, MacKaye stated in his address, "The Appalachian Trail is a footway and not a motorway because it is a primeval way and not a metropolitan way."[101]

Although MacKaye held these views of the Appalachian Trail, not all of the members of the Potomac Appalachian Trail Club agreed with him. While all of the members believed in the construction of the Appalachian Trail, there existed in the PATC membership a reasonably powerful group that believed in compromising with the Skyline Drive boosters in order to get the AT completed. This conciliatory group had at least as much influence on the nature of the Drive and the parkway in the region as MacKaye and others who were less willing to compromise.

4.1.1 The Varied Stance of the PATC and a Completed Appalachian Trail

A segment of the Potomac Appalachian Trail Club membership held out against the Skyline Drive, while another worked with the Park Service and the builders of the Skyline Drive and actually improved the surface quality of the Trail in certain places. During the New Deal era, the Civilian Conservation Corps rebuilt, to a higher standard, sections obliterated by the Drive.[102]

For the most part, PATC members were middle-class professionals, "whose vocations and activities for the trail were largely separate"[103] from their professional lives. The middle-class membership of the PATC (and other Trail clubs) built the Trail, according to geographer Ronald Foresta, as "an alternative to nonconstructive urban leisure."[104] Indeed most were not as interested in the longer-term social reform and regional planning aspects of Benton MacKaye's 1921 article. Foresta argues that the Trail idea was transformed from a social reform movement modeled after MacKaye's thinking into a place for individual recreation as conceived by the middle class professionals who built the Trail.[105]

Pragmatists from the PATC intent on completing the Trail, like Myron H. Avery, the PATC's President during this period,[106] not only refused to embrace MacKaye's social reform vision, but went one step further, promoting the completion of the Trail instead of insisting that it be built within a purely primeval environment.[107] MacKaye himself wrote in the early 1921 "Memorandum: Social Planning and Social Readjustment," that recreational aspects of the Appalachian Trail would have to precede any further social reform.[108] And by the late 1920s, it's arguable that MacKaye's vision of the Trail, at least the section in the Shenandoah National Park, had progressed to the point that his social reform agenda had been dropped. That is, MacKaye supported the Park in his 1925 letter to Harlan Kelsey. Development within National Park boundaries along the lines of MacKaye's Appalachian Trail proposal[109] was nonetheless development and excluded from National Parks according to public law. By 1930, the vision of the Appalachian Trail, according to MacKaye's address to the Appalachian Trail Conference held at Skyland, was a footpath through the primeval, not a motorway.[110] He summed up that address with the exhortation, "To preserve the primeval environment: this is the point, the whole point and nothing but the point, of the Appalachian Trail."[111]

MacKaye and the anti-road "holdouts" did compromise up to a point. In 1934, MacKaye published "Flankline vs. Skyline," which represented MacKaye's shifting understanding of the automobile as recreation in and of itself. MacKaye, using the Skyline Drive as an example, argued that the appropriate route for a motorway was the Flankline. This line preserved the ridgeline for the serious hiker and reserved the primeval for the foot traveler. As a place for the automobile, it also allowed for better views and a variety of scenery. That is, MacKaye proposed that the Flankline route periodically climb to the ridgeline and afford the motorist an appropriately majestic view from the mountain.[112] In this piece MacKaye acceded to the idea that the motorist could actually participate in recreation during his or her drive.

The conflict over skyline roads and the Appalachian Trail among Appalachian Trail Conference membership came to a head at the June 1935 ATC meeting held at Skyland. MacKaye and Avery made their points. MacKaye's was an either/or position: either a truly wild Trail (an absence of interfering motorways), or no Trail at all. Avery supported the idea that Trail builders needed to work with the appropriate government agencies to complete the Trail and then relocate sections of it eliminated by skyline roads.[113]

The Appalachian Trail Conference adopted a resolution in June 1935 that stated in essence that the Conference would work with the appropriate officials to relocate the Appalachian Trail where current road projects interfered with it. With references to future projects, the Appalachian Trail Conference resolved to study each case on its own merits. Included among those cases were the southern-most section of the Skyline Drive (Swift Run Gap to Jarman's Gap), the proposed Park-to-Park Highway (later called the Blue Ridge Parkway), and others such as the Green Mountain Parkway (see the next chapter).[114] Not surprisingly, MacKaye labeled the resolution "meaningless" in a letter to Avery later that year.[115]

Avery lived up to the resolution's call that "each project should be considered on its particular merits."[116] In early 1936, Avery presented Secretary of the Interior Harold L. Ickes with a plan to run the southern third of the Skyline Drive along the western flank of the Blue Ridge, thereby continuing the Drive without bisecting the southern section of the Park. This route, Avery wrote, would "through detouring the Skyline Drive through the southern portion of the Shenandoah National Park [. . .] leave this area in its present condition, untraversed by the road."[117] Ickes sent the proposal to Thomas Vint and J.R. Lassiter, the superintendent of Shenandoah National Park. Although Avery's proposal was not accepted (Vint and Lassiter cited flood concerns, the need for additional Park entrances, and increased costs from additional bridges and maintenance),[118] his audience with the Department of the Interior and the Park Service showed that his positions were at the least seriously considered. Indeed, the Blue Ridge Parkway, running from the southern terminus of the Shenandoah National Park to the Great Smoky Mountains National Park, does not continuously follow the ridgeline, thus ultimately reflecting Avery's pragmatic positions.

4.2 The Park Boosters – Metropolitanist Motivations and Actions

The Shenandoah National Park and Skyline Drive boosters consistently pushed for road development as a means of regional economic development. These boosters, all park supporters by virtue of their membership in the local chambers of commerce, regional business concerns like Shenandoah Valley Incorporated,[119] and the Virginia Commission on Conservation and Development, wanted economic development in Virginia. Had the opportunity arisen, this development could have taken the form of industry and manufacturing concerns, a federal initiative such as a military installation, or any number of other types of projects.[120]

The boosters of the Park and Drive were uniform in their view that the maximization of each of the inherent resources would be an economic benefit to Virginia and the localities therein. The immediate influx of federal dollars, as exemplified by the "drought relief" funds of 1931, brought benefits to a region. Roads – infrastructure – would bring people to Virginia just as infrastructure had brought people to New York or Boston, and, the boosters asserted, those people would spend their money in Virginia. Virginia had, at least conceptually, the promise of the natural environment in the Blue Ridge Mountains, a cultural and historical heritage accessible to nearly every American who knew at least something about American history, and the political will to follow the dominant economic paradigm. In the eyes of the economic development boosters, there was no questioning the idea that bringing a tourism-based economic model (metropolitan in origin) into the most remote and rural areas of the state was the right thing to do.[121]

Surveys of the Park area did not include an assessment of the human resources (in other words, the underrepresented, disenfranchised mountain people) or recognition of the rural or regional economy and culture.[122] These resources and the economic system that supported them were outside of the dominant economic paradigm.[123] Without making clear, calculated value judgments, Virginia officials embarked on a policy to replace the existing, however outdated, economic system with one that readily benefited the middle and upper classes. Virginia officials saw only the statewide economic benefits. For its part, the National Park Service gained a resource within a day's drive of millions of Americans. This grew the Park Service's middle-class constituency almost immediately. However, the pursuit of this development by the boosters overwhelmed the region and robbed it of the very uniqueness that made it appealing in the first place.

4.2.1 The Earliest Park and Drive Boosters

In early 1924, Secretary of the Interior Hubert Work established the Southern Appalachian National Park Commission (SANPC). The Commission's mandate spurred on two groups that had taken an early interest in developing a park in the Blue Ridge Mountains of Virginia: the regional business group Shenandoah Valley Incorporated, and a looser association headed by George Freeman Pollack, proprietor of Skyland. By the fall of 1924, these two groups had merged together as the Northern Virginia National Park Association. This organization lobbied the members of the SANPC by means of organized outings at Skyland and general publicity. By December 1924, the SANPC had recommended that the Shenandoah area be considered for a national park.[124] By the summer of 1925, the Northern Virginia National Park Association combined forces with the Virginia State Chamber of Commerce to form the Shenandoah National Park Association.[125] It was at this point that a coordinated statewide and national fundraising and publicity campaign officially began.

Dr. W.J. Showalter, formerly Editor of the Harrisonburg *Daily News Record* and, by the middle part of the 1920s, an Associate Editor of *National Geographic*,[126] worked tirelessly to tout the economic benefits of the park. In 1926, he argued in the *Washington Post* that the proposed park would bring economic benefits to not only Virginia's Shenandoah Valley and Piedmont, but also to Washington, DC.[127] Citing the SANPC report that noted that 40,000,000 people lived within a day's ride of the Blue Ridge, Showalter estimated that 750,000 persons would annually visit the Park and at least 200,000 of those would stop in the Nation's capital. Inevitable economic prosperity for Virginia was obvious to Showalter, who argued that "buying power [meant] an increase in your [Washington] business."[128]

The likelihood and allure of the park during the years following the SANPC report led to the establishment of numerous business organizations intent on benefiting from the park's creation. Chief among these were road and highway organizations. Most often, these groups grew out of the associations between county boards of supervisors, local chambers of commerce, and other business interests. They took names such as the Shenandoah National Park Circuit Highway Association, the Inter-County Lee Highway Committee, the John Marshall Highway Association, and the Washington Boulevard Association.[129]

Each of these groups believed in and furthered the idea that the development of the Park would bring economic prosperity to the region. They appealed to prospective supporters by touting the possibly unique claim of someday profiting from the "East's First Great Park."[130] These groups also emerged as the foundation for an automobile-based constituency that would push for the use of the automobile once the Park Service began to develop Shenandoah's resources. Certainly, the highway and road associations only mirrored others established during this period, as the use of the automobile became more widespread and available to the middle class. The importance of the automobile in the pursuit of commercial development was not lost on any business interest. The pursuit of the automobile tourist dollar literally "drove" the constitutions of these groups. Ferdinand Zerkel, associated with many of the business groups beginning in 1924, recounted the attitude of the business community during this early period in a 1931 letter:

> I send you some 1924 and 1925 clippings and booklets that I hope, at least in the marked sections, will prove that the State of Virginia became interested in the Park largely on the commercial prospects for all the state.[131]

In the minds of virtually all booster groups, then, economic development was predicated on tourist dollars and tourist dollars were to flow into Virginia via the automobile. W.J. Showalter noted this in his April 1926 article in the *Washington Post*,[132] as did nearly every other Park enthusiast. When Roy Lyman Sexton had begun to make an issue over the Skyline Drive's impact on the Appalachian Trail, Zerkel responded by writing, "Again, the Trail is incidental and those who have been 'Park Nuts' consistently for six and a half years [know this]."[133] A Shenandoah National Park Association publication in the middle 1920s may have summed up the prospect for development. It read:

> Inquiry among superintendents of the great western parks has shown, too, a singular psychology. Tourists visiting the parks, whether by motor or train are never content to see only the park area and then hasten home. They seldom fail to make a general tour of the entire State. In the State of Washington, for instance, tourists left hundreds of thousands of dollars in Mt. Rainier National Park proper. But by actual count they left millions in the State at large, State officials discovered.[134]

4.2.2 The State Commission on Conservation and Development Efforts

The many and varied groups interested in promoting the Park had neither the political clout nor the organizational structure to manage the entire effort. Soon after being inaugurated in 1926, Virginia Governor Harry F. Byrd appointed William E. Carson Chairman of the SCCD, which assumed control of Virginia's effort to establish the park. The Commission formalized the State's efforts at advocating the economic benefits of park development. The Commission's purpose of developing the recreational, scenic, and historic resources available in Virginia fit within the framework of the business climate that surrounded the early history of the Park. Writing in late 1933, W.J. Showalter suggested that if not for the Commission, there would never have been a Shenandoah National Park.[135]

From the earliest meetings, the prospect of tourist dollars and the complementary development drove the Commission's policies.[136] The Commission worked toward a policy that it believed would allow Virginia to keep up with the business advances available to other states. Whereas manufacturing and industry may have been appropriate elsewhere, the Commission believed in emphasizing Virginia's natural and historic resources.[137]

Carson's Commission, by 1933, had facilitated construction of the Skyline Drive, the Shenandoah National Park, and the Colonial National Monument. Additionally, this commission had developed President Hoover's Camp, other historic sites, waterpower resources, state parks and forests, and supervised the work of the State Geological Survey.[138] With Shenandoah National Park enthusiasts, Carson tried to channel the energy toward the goal of getting the Park established as soon as possible. Indeed, his work in getting the drought-relief funds available for the first section of the Skyline Drive effectively saved the entire project.[139] He also wrote the plan by which Virginia would initially turn over deeds only to the land needed to build the Skyline Drive

Carson and the Virginia Commission on Conservation and Development certainly succeeded in effecting the policy of promoting Virginia's economy through tourism dollars. By 1935, $73 million per year was flowing into Virginia via the tourist trade, equaling the industrial sector. Gasoline tax receipts were the sixth highest in the country, second in the South only to Florida.[140]

4.2.3 The Booster Legacy

Virginia's economy had, in part at least, evolved into an automobile-based tourist economy by 1935. It wasn't long before every Chamber of Commerce, Board of Supervisors, and local business organization recognized this. Throughout the early stages of the Shenandoah National Park movement, the prospect of 40 million people within a day's drive motivated the public actions of these private groups. These groups lobbied the Park Service and the SCCD for their fair share of the pie in the development of the State's resources. The business organization that supported the John Marshall Highway Association – a group promoting a "coast to coast" highway – worked to have the northern-most section of the Skyline Drive intersect with the proposed John Marshall Highway. The city of Waynesboro and its business and community leaders tried, to no avail, to have the Skyline Drive cross their municipal boundary. Falling short of that, they lobbied Kentucky Congressman Maurice K. Thatcher to have the National Park-to-Park Highway run through their city limits. When that did not work out, they aimed at getting the Blue Ridge Parkway.[141]

However diligently these groups worked to gain additional business opportunities through the construction of roads, their most significant accomplishment had been achieved by the time the first survey of the Skyline Drive had been completed. By then, they had established themselves as a formidable constituency – automobile-oriented and supportive of the dominant economic paradigm, as opposed to the reform-minded ideology that MacKaye had pushed for the region. Over the short term, the fact that Waynesboro did not get the Skyline Drive, the Blue Ridge Parkway, or the National Park-To-Park Highway hardly represented the lost opportunity of a lifetime. The tourist economy changed radically in the long term, and communities like Waynesboro that lost out initially, as well as the early "winners," were left behind.[142] Ironically, as tourism in the United States evolved and people were more willing to drive longer distances to vacation destinations, localities near the Shenandoah National Park changed their emphasis. The state highway once supposed to bear the names "Shenandoah National Park Circuit Highway" and "Lee Highway" is now called "Seminole Trail" in an effort to attract the automobile tourists traveling from the northeast down to Florida.

5. CONCLUSION

The history of the Skyline Drive and the Shenandoah National Park reveals the presence of two influential yet limited forces that pressured, twisted, and tweaked the planning, design, and implementation of the Drive and the Park. Although the federal government had a significant role – through the allocation of relief funds, the Southern Appalachian Regional Park Commission, the Park Service's desire to expand its constituency, and other development projects in and around the Park – the Skyline Drive was certainly not born of federal policy. For their part, the Park and road boosters viewed the Skyline Drive and numerous other roads in the region as facilitators of economic growth. The pursuit of this economic development followed (perhaps a little too slavishly) the dominant economic themes of the first part of this century.

While Lewis Mumford and Benton MacKaye may not have understood or conceptualized the best way out of the economic circumstances associated with the Fourth Migration – Regional Planning – they certainly understood the patterns of decentralization and metropolitanization and the inefficient uses of resources. The Skyline Drive boosters meanwhile embraced this economic pattern with open arms (if not open eyes) and saw within it the path to economic prosperity. The introduction of tourism and metropolitanizing influences into the Blue Ridge Mountains, Central Virginia, and the Shenandoah Valley was all the boosters could ask for. The regional visionaries, on the other hand, saw the Blue Ridge Mountains (and more broadly the Appalachian chain), as a natural dam against urbanization. Although the effort to preserve the wilderness along the ridgeline only rarely reflected Benton MacKaye's arguments, his influence was clearly evident in arguments made in favor of wilderness. MacKaye and the other Potomac Appalachian Trail Club members who truly believed in his vision fought for wilderness to the end, despite MacKaye's knowledge that an all-volunteer workforce caught up in construction would not, by its very nature, subscribe to the social-reform aspects. During the 1920s, MacKaye had not yet completely abandoned the notion of the communal work camps along the Trail. His stubborn insistence upon clinging to the ridgeline as the only site for a wilderness section raises an interesting issue. Never suggested was a compromise in which the flanks of the mountains would be set aside as the wilderness corridor. I would argue that this could not occur because, again, although MacKaye later gave up on the social-reform issues he felt could be served by the Trail, he never gave up on the idea of observing the region from the top of the ridgeline. And, further, he would not surrender the ridge to the forces of metropolitanization.

At a minimum, the Appalachian Trail of Benton MacKaye's vision was outside the dominant economic paradigm. The development of MacKaye's wilderness Appalachian Trail as a trail alone served as a model for the Regional Planning Association's Fourth Migration. Meanwhile, the Trail as conceived by Myron Avery prioritized completion and a connected trail over the MacKaye vision. Because the PATC membership did not fully embrace the MacKaye vision, the vacuum created by MacKaye's refusal to compromise perhaps opened the way for the metropolitanization of the region (or at least the diminishing of its cultural uniqueness) with the construction of the Skyline Drive.

---

[1] Southern Appalachian National Park Commission, *Final Report of the Southern Appalachian National Park Commission to The Secretary of the Interior, June 30, 1931* (Washington, DC: GPO, 1931) 8.

[2] Benton MacKaye, "Re Skyline Drives and the Appalachian Trail," MacKaye Family Papers, Dartmouth College Library, Hanover, NH.

[3] Benton MacKaye, "An Appalachian Trail: A Project in Regional Planning," *Journal of the American Institute of Architects* 9 (1921): 325-330.
[4] The Potomac Appalachian Trail Club did not take a unified stand against the Skyline Drive throughout the entire early history of the Club and the Drive.
[5] Harvey Broome, "Origins of the Wilderness Society," *The Living Wilderness* 5 (1940): 13.
[6] Benton MacKaye, "An Appalachian Trail" 325-330.
[7] Broome 13.
[8] Charles W. Eliot, 2nd, "The Influence of the Automobile on the Design of Park Roads," *Landscape Architecture Magazine* 13 (1922): 27ff.
[9] Ferdinand Zerkel to Roy Lyman Sexton, Feb. 14, 1931, Ferdinand Zerkel Papers, Shenandoah National Park Archives, Luray, VA.
[10] Charles Perdue and Nancy J. Martin-Perdue, "Appalachian Fables and Facts: A Case Study of the Shenandoah National Park Removals," *Appalachian Journal* 7:1-2 (1979-1980): 84-105. See especially page 89 and note 29.
[11] Perdue and Martin-Perdue, "Appalachian Fables," 89 and note 29.
[12] Gene Wilhelm Jr., "Shenandoah Resettlements," *Pioneer America* 14:1 (March 1982): 15-41.
[13] Charles Perdue and Nancy J. Martin-Perdue, "'To Build a Wall Around These Mountains': The Displaced People of Shenandoah," *The Magazine of Virginia History* 49 (1991): 50.
[14] Lewis Mumford, "Regions – To Live In," *The Survey Graphic* 54 (1925): 151-152.
[15] Southern Appalachian Regional Park Commission, *Final Report* 7-8.
[16] Other members appointed to the SANPC included Congressman Harry W. Temple, industrialist William C. Gregg, Colonel Glenn S. Smith of the United States Geological Society, and Major William A. Welch, manager of the Palisades Interstate Park.
[17] Darwin Lambert, *The Undying Past of the Shenandoah National Park* (Boulder, CO: Roberts Rinehart, Inc., 1989) 200, and also Harlan P. Kelsey to Barrington Moore, March 27, 1925, National Parks: Shenandoah; Central Classified Files 1907-1942, 1907-1932, File 0.32; National Archives at College Park, MD (NACP).
[18] This basic chronology comes from a number of sources. The most useful has been Darwin Lambert, "Shenandoah National Park, Administrative History, 1924-1976" (Luray, VA, 1979; unpublished photocopy) 1, and William E. Carson, *Conservation and Development in Virginia: Outline of the Work of the Virginia State Commission on Conservation and Development* (Richmond, VA: Division of Purchase and Printing, 1934) 3-4. See also Dennis Simmons, "Conservation, Cooperation, and Controversy: The Establishment of Shenandoah National Park, 1924-1936," *Virginia Magazine of History and Biography* 89 (1981): 387-404.
[19] Simmons, "Conservation, Cooperation, and Controversy" 392-395.
[20] Simmons, "Conservation, Cooperation, and Controversy" 396-397. See also J.M. Samuels to Governor Harry F. Byrd, April 29, 1927, Harry F. Byrd, Executive Papers, 1926-1930. Virginia State Library, Richmond, VA. Samuels, as Secretary of the Orange County Chamber of Commerce, was asking Byrd to figure out a way to get public money appropriated to the park fund. North Carolina and Tennessee had floated bonds to add money to their park funds. Virginia law prevented the State from floating bonds for such a purpose.
[21] Hubert Work, Secretary of the Interior, to Governor Harry F. Byrd, December 22, 1927. General Files; Central Classified Files 1907-1942, 1907-1932, File 0.32-101; Records of the National Park Service, Record Group 79; NACP.

---

[22] Lambert, "Administrative History," for the chronology. See also Simmons, "Conservation, Cooperation, and Controversy," 395. Simmons pointed out that the original assessment of the money needed to buy the nearly 400,000 acres had been made by adding up all the locally assessed values of all the parcels within the intended park boundary. Markct value outpaced the assessed value of the aggregated properties nearly three-fold.

[23] Hubert Work to Harry F. Byrd, December 22, 1927; General Files; Central Classified Files 1907-1942, 1907-1932, 0.32-101; NPS, RG 79; NACP.

[24] Arno B. Cammerer, *Report to the Secretary of the Interior on the Boundaries of the Shenandoah National Park as Provided under the Act of Congress Approved May 22, 1926*, December 21, 1927; General Files; Central Classified Files 1907-1942, 1907-1932, 0.32-101; NPS, RG 79; NACP.

[25] Cammerer, *Report to the Secretary of the Interior.*

[26] Robert Sterling Yard, ed., "Shenandoah National Park Named by Committee," *National Parks Bulletin* 42 (December 24, 1924). Copy available at the National Archives. National Parks: Shenandoah; Central Classified Files 1907-1942, 1907-1932, File 0.32 ; Records of the National Park Service, Record Group 79; NACP.

[27] Dennis E. Simmons, "The Creation of Shenandoah National Park and the Skyline Drive, 1924-1936," diss., U of Virginia, 1978, 63-66.

[28] Simmons, "The Creation of Shenandoah," and also Simmons, "Conservation, Cooperation, and Controversy" 397.

[29] See Lambert, "Administrative History" and Sarah Georgia Harrison, "The Skyline Drive: A Western Park Road in the East," *Parkways: Past, Present, and Future* (Boonc, NC: Appalachian Consortium P, 1987) 38-48.

[30] Dennis E. Simmons, "The Creation of Shenandoah" 70-78.

[31] Arthur Davidson, "Skyline Drivc and How it Came to Virginia," Conserving and Developing Virginia. *Report of W.E. Carson, Chairman, State Commission on Conservation and Development, July 26, 1926 to Dec. 31, 1934* (Richmond: Division of Purchasing and Printing, 1935) 75-79.

[32] See William E. Carson to Colonel Glenn Smith, October 7, 1929, General Files; Central Classified Files 1907-1942, 1907-1932, File 0.32-101; NPS, RG 79; NACP; and Simmons, "The Creation of Shenandoah National Park" 7ff; and Davidson 77.

[33] Simmons, "The Creation of the Shenandoah" 75-77; and William E. Carson to Horace M. Albright, November 3, 1930, Central Classified Files 1907-1942, 1907-1932, File 631.1-901; Records of the National Park Service, Record Group 79; NACP.

[34] Thomas Vint to Horace Albright, *Memorandum to the Director*, March 16,1931, Central Classified Files 1907-1942, 1907-1932, File 631.1-901; Records of the National Park Service, Record Group 79; NACP.

[35] William E. Carson to Louis C. Cramton, December 31, 1930, Central Classified Files 1907-1942, 1907-1932, File 630; Records of the National Park Service, Record Group 79; NACP.

[36] Horace M. Albright to Carter Glass, December 11, 1930, Central Classified Files 1907-1942, 1907-1932, File 0-32; Records of the National Park Service, Record Group 79; NACP; William E. Carson to Horace M. Albright, November 30, 1930, Central Classified Files 1907-1942, 1907-1932, File 631.1-901; Records of the National Park Service, Record Group 79; NACP; and Arno B. Cammerer to Kenneth McKellar, Central Classified Files 1907-1942, 1907-1932, File 630; Records of the National Park Service, Record Group 79; NACP.

[37] U.S. Department of the Interior, *Memorandum*, Feb. 4, 1931, General Files; Central Classified Files 1907-1942, 1907-1932, National Parks: Shenandoah, File 0.32-120 ; NPS, RG 79; NACP.

[38] William E. Carson to Louis C. Cramton, December 31, 1930, Central Classified Files 1907-1942, 1907-1932, File 630; Records of the National Park Service, Record Group 79; NACP.

[39] Carson to Cramton, December 31, 1930.

[40] See Walter W. Mallonee, *Origin of the Skyline Drive Through the Shenandoah National Park in the Blue Ridge Mountains of Virginia* (United States: W.W. Mallonee, 1995) 2-3. Walter Mallonee was employed by the Bureau of Public Roads beginning in January 1931.

[41] Charles E. Peterson to Horace M. Albright, "Memorandum to the Director," April 14, 1931, Central Classified Files 1907-1942, 1907-1932, File 631.1-901; Records of the National Park Service, Record Group 79; NACP.

[42] Roy Lyman Sexton to William E. Carson, December 22, 1930, Central Classified Files 1907-1942, 1907-1932, File 630; Records of the National Park Service, Record Group 79; NACP; George Freeman Pollack to William E. Carson, November 20, 1930, Central Classified Files 1907-1942, 1907-1932, File 631.1-901; Records of the National Park Service, Record Group 79; NACP; and Ferdinand Zerkel to William E. Carson, December 5, 1930, Central Classified Files 1907-1942, 1907-1932, File 631.1-901; Records of the National Park Service, Record Group 79; NACP.

[43] U.S. Department of the Interior, *Memorandum*, Feb. 4, 1931.

[44] "Seph" Remine, letter, *Knoxville News-Sentinel*, May 3, 1931, clipping in Central Classified Files 1907-1942, 1907-1932, File 630; Records of the National Park Service, Record Group 79; NACP.

[45] Editor's response to "Seph" Remine, letter, *News-Sentinel*, May 3, 1931, clipping in Central Classified Files 1907-1942, 1907-1932, File 630; Records of the National Park Service, Record Group 79; NACP.

[46] Horace M. Albright to Edward J. Neeman, May 7, 1931, Central Classified Files 1907-1942, 1907-1932, File 630; Records of the National Park Service, Record Group 79; NACP.

[47] Roy Lyman Wilbur to J. Will Taylor, May 13, 1931, Central Classified Files 1907-1942, 1907-1932, File 630; Records of the National Park Service, Record Group 79; NACP.

[48] Arno B. Cammerer to Kenneth McKellar, May 13, 1931, Central Classified Files 1907-1942, 1907-1932, File 630; Records of the National Park Service, Record Group 79; NACP.

[49] Kenneth McKellar to Arno B. Cammerer, May 15, 1931, Central Classified Files 1907-1942, 1907-1932, File 630; Records of the National Park Service, Record Group 79; NACP.

[50] Arno B. Cammerer to Kenneth McKellar, May 20, 1931, Central Classified Files 1907-1942, 1907-1932, File 630; Records of the National Park Service, Record Group 79; NACP.

[51] Editor's response to "Seph" Remine, letter.

[52] See Lambert, "Administrative History" 100-110; Davidson 75-79.

[53] U.S. Department of the Interior, *Memorandum*, Feb. 4, 1931.

[54] U.S. Department of the Interior, *Memorandum for the Press*, 25 March 1931, General Files; Central Classified Files 1907-1942, 1907-1932, National Parks: Shenandoah File 630; NPS, RG 79; NACP.

[55] Mallonee, *Origin of the Skyline Drive* 2-3.

---

[56] L. Ferdinand Zerkel to William E. Carson, January 26, 1931, L. Ferdinand Zerkel Papers, Shenandoah National Park Archives, Luray, VA.

[57] Arno B. Cammerer to L. Ferdinand Zerkel, February 7, 1931, and Cammerer to Zerkel, February 16, 1931, L. Ferdinand Zerkel Papers, Shenandoah National Park Archives, Luray, VA. Two things are of interest here: first, by including at least two extremely prominent members of the PATC, the Park Service officials were likely trying to build support for the road with a group that was founded on the idea of supporting the Park despite the notion that the road would disturb their work on the Trail – Zerkel had already recommended to Carson that the route of the road follow the Appalachian Trail in certain places (see Zerkel to Carson, January 26, 1931, L. Ferdinand Zerkel Papers, Shenandoah National Park Archives, Luray, VA); and second, Cammerer may have spoken of the unsettled nature of the road in anticipation of further political problems such as the one that occurred with Senator McKellar only two months later.

[58] Zerkel to Cammerer, Feb. 14, 1931, L. Ferdinand Zerkel Papers, Shenandoah National Park Archives, Luray, VA. Cammerer responded on February 16 and reminded Zerkel that the PATC members were friends of the Park movement and should be treated as such. This exchange is indicative of the growing concern within the PATC over the road project. Zerkel's discussion of the PATC concern over roads is extremely informative with regard to the activist boosters and their desire for roads as catalysts for economic development.

[59] Thomas Vint to Horace Albright, Memorandum to the Director, March 16, 1931, Central Classified Files 1907-1942, 1907-1932, File 631.1-901; Records of the National Park Service, Record Group 79; NACP.

[60] Vint to Albright, March 16, 1931. Vint wrote his memo in March 1931 when the minimum park acreage remained at 327,000 acres. Ultimately, the State of Virginia gained authority to reduce the minimum acreage to 160,000 acres.

[61] U.S. Department of the Interior, *Memorandum for the Press*, 25 March 1931.

[62] See Charles E. Peterson to A.E. Demaray, July 27, 1931, Central Classified Files 1907-1942, 1907-1932, File 631.1-901; Records of the National Park Service, Record Group 79; NACP.

[63] A.E. Demaray to H.K. Bishop, March 28, 1931, and Bishop to Demaray, March 31, 1931, Central Classified Files 1907-1942, 1907-1932, File 631.1-901; Records of the National Park Service, Record Group 79; NACP.

[64] Gilmore Clarke to A.E. Demaray, July 6, 1931, Central Classified Files 1907-1942, 1907-1932, File 631.1-901; Records of the National Park Service, Record Group 79; NACP.

[65] Linda Flint McClelland, *Presenting Nature: The Historic Landscape Design of the National Park Service, 1916-1942* (Washington, DC: Government Printing Office, 1994) 104.

[66] Charles E. Peterson to Horace M. Albright, July 8, 1931, Central Classified Files 1907-1942, 1907-1932, File 631.1 – 901; Records of the National Park Service, Record Group 79; NACP.

[67] Peterson to Albright, July 8, 1931. Peterson continued in his letter, "since we have had only a vague silence with the [two unreadable words] that C & D have been to see Mr. Rockefeller, and vice versa, etc... [sic] This makes me sweat for fear that the Park Service is going to lose out on a chance to have a really distinguished project – what I mean is one with some guts in it."

[68] Charles E. Peterson to A.E. Demaray, July 27, 1931, Central Classified Files 1907-1942, 1907-1932, File 631.1-901; Records of the National Park Service, Record Group 79; NACP.

[69] Peterson to Demaray, July 27, 1931.

[70] Peterson to Demaray, July 27, 1931.
[71] Peterson to Demaray, July 27, 1931.
[72] McClelland 134-135.
[73] Gilmore Clarke to A.E. Demaray, September 14, 1931, Thomas MacDonald to A.E. Demaray, August 29, 1931, Central Classified Files 1907-1942, 1907-1932, File 631.1-901; Records of the National Park Service, Record Group 79; NACP.
[74] McClelland 135.
[75] See McClelland.
[76] A.E. Demaray to George Moskey, April 4, 1931, General Files; Central Classified Files 1907-1942, 1907-1932, National Parks: Shenandoah File 630; NPS, RG 79; NACP.
[77] Lambert, "Administrative History" 107.
[78] Perdue and Martin-Perdue, "'To Build a Wall Around These Mountains" 48ff.
[79] Arno B. Cammerer to Horace M. Albright, "Memorandum for the Director," November 4, 1932, General Files; Central Classified Files 1907-1942, 1907-1932, National Parks: Shenandoah File 304-611; NPS, RG 79; NACP.
[80] See, for example, Benton MacKaye, "Some Social Aspects of Forest Management," *Journal of Forestry* 16 (1918): 210-214.
[81] Harold Anderson, "What Price Skyline Drives?" *Appalachia* 21 (1935): 410ff.
[82] Harold Allen and L.F. Schmeckebier, "Shenandoah National Park: The Skyline Drive and the Appalachian Trail," *Appalachia* 22 (1936): 76ff.
[83] Allen and Schmeckebier 76-78.
[84] See Lambert, "Administrative History" 109.
[85] Davidson 75-79
[86] See Lambert, "Administrative History" 3.
[87] Zerkel to Cammerer, Feb. 14, 1931, and Harold C. Allen to Ferdinand Zerkel, April 4, 1931, L. Ferdinand Zerkel Papers, Shenandoah National Park Archives, Luray, VA.
[88] Harlan P. Kelsey to Benton MacKaye, March 24, 1925, also, MacKaye to Kelsey, March 18, 1925, MacKaye Family Papers, Dartmouth College Library, Hanover, NH.
[89] Kelsey to MacKaye, March 27, 1925, MacKaye Papers.
[90] Kelsey to MacKaye, March 27, 1925, MacKaye Papers.
[91] From the December 12, 1924 preliminary report published in the Southern Appalachian National Park Commission, *Final Report* 8.
[92] MacKaye to Kelsey, March 31, 1925, MacKaye Papers.
[93] Harlan P. Kelsey to Barrington Moore, March 27, 1925, National Parks: Shenandoah; Central Classified Files 1907-1942, 1907-1932, File 0.32; Records of the National Park Service, Record Group 79; NACP. Note Kelsey's concern with the road bisecting a 10-to-12-mile-wide park.
[94] Horace M. Albright to Harlan P. Kelsey, April 17, 1931, Central Classified Files 1907-1942, 1907-1932, File 631.1-901; Records of the National Park Service, Record Group 79; NACP.
[95] Benton MacKaye to Arno B. Cammerer, June 13, 1932, Central Classified Files 1907-1942, 1907-1932, File 631.1-901; Records of the National Park Service, Record Group 79; NACP. See also Harold C. Anderson to Benton MacKaye, June 10, 1932, MacKaye Papers.
[96] MacKaye to Cammerer, June 13, 1932.
[97] MacKaye to Cammerer, June 13, 1932.

---

[98] Harlan P. Kelsey to Horace M. Albright, November 22, 1925, Central Classified Files 1907-1942, 1907-1932, File 631.1-901; Records of the National Park Service, Record Group 79; NACP.

[99] Benton MacKaye, "The 'Open Door' in America: Manifest Destiny for Appalachia, Not Manchuria," unpublished, 1932, MacKaye Papers.

[100] MacKaye, "The 'Open Door' in America."

[101] Benton MacKaye, "Vision & Reality," Address to the Appalachian Trail Conference, May 30, 1930, Skyland, VA, MacKaye Papers.

[102] See Anderson, "What Price Skyline Drives?" Anderson wrote the following about the parts of the Appalachian Trail reconstructed by the Park Service and the CCC: "The 'AT' marker which formerly led one over a primitive footpath or over old wood roads now leads one over a beautifully graded and banked horseback trail. Some consider this a 'super-trail' and a vast improvement over the old rough, rugged, in some places steep, and oftentimes more or less overgrown footpath. Lovers of the wild and primitive, however, consider this new thoroughfare a far cry from the Appalachian Trail as originally conceived, for the new trail possesses little of wilderness characteristics."

[103] Ronald Foresta, "Transformation of the Appalachian Trail," *Geographical Review* 77 (1987): 79.

[104] Foresta 84.

[105] Foresta 79.

[106] Foresta wrote of Avery, "[He] more than anyone else was responsible for the success of the trail." Foresta 79.

[107] MacKaye and Avery had a heated exchange of views – at the Seventh Appalachian Trail Conference held at Skyland in June 1935 and in a series of letters, MacKaye to Avery, November 20, 1935, and Avery to MacKaye, December 19, 1935, MacKaye Papers.

[108] Benton MacKaye, "Memorandum: Social Planning and Social Readjustment," undated memorandum (likely 1921), MacKaye Papers. Foresta argued that the Appalachian Trail could never achieve the social reform goals set out by Benton MacKaye because the construction was left to middle class professionals, not the working class whom MacKaye had intended the reforms to benefit. Historian Paul Sutter pointed out that this analysis was wrong given MacKaye's discussion of the Trail and the complementary social reform communities intended to follow the construction of the Trail. See Paul Sutter, "Labor and Natural Resources: Colonization, the Appalachian Trail, and the Social Roots of Benton MacKaye's Wilderness Advocacy," presented at the University of Virginia History Department Seminar. Charlottesville, VA, September 28, 1998; unpublished manuscript in the possession of the author. MacKaye knew that the Trail would have to be constructed by volunteers who would primarily be professionals from the middle class. While he expected that some would understand the value of remaining year round outside the cities along the Trail, he certainly knew that the entire concept could be co-opted by those volunteers as a recreational space for and of the middle class. I think this in part explains MacKaye's fight for the wildness of the Trail over its social reform aspects.

[109] See Chapter 3, specifically MacKaye, "An Appalachian Trail," and MacKaye, "Memorandum: Social Planning and Social Readjustment."

[110] MacKaye, "Vision & Reality."

[111] MacKaye, "Vision & Reality."

[112] Benton MacKaye, "Flankline vs. Skyline," *Appalachia* 20 (1934): 104-108. In a letter to Arno Cammerer, then the Director of the National Park Service, MacKaye defended motorists, citing "Flankline vs. Skyline," but, he noted that the wilderness required

separation from the motorist. He wrote, "[. . .] wilderness and not an outing area. Some people want Coney Island: let them have it in an outing area. Other people want the opposite of Coney Island: they should have it in a wilderness area." See MacKaye to Arno B. Cammerer, September 21, 1934, MacKaye Papers.

[113] See E.A. Dench, "Hiking and Camping Forum." *Nature Magazine* 26(1935): 186-188. Dench's criticism of the Trail's close proximity to the Skyline Drive mirrored that of Anderson cited above. On the other hand, Dench noted the fact that the blue markers of the Appalachian Trail did open a "gateway" to the outdoors.

[114] See the "Resolution" of the Appalachian Trail Conference, June 22-23, 1935, Shenandoah National Park Papers, Archives, Shenandoah National Park Headquarters, Luray, VA.

[115] MacKaye to Avery, 20 November, 1935, MacKaye Papers. In this letter, MacKaye reminded Avery of his own words to Arno B. Cammerer regarding the Park-to-Park Highway project which he referred to as the "(Skyline) Road project." On August 7, 1934, Avery had written the following to Cammerer: "There is much in the (Skyline) Road project which could be questioned. I mention the economic factors of high cost, difficult construction, prevalence of fog and occasional snow, surfeit of scenery, a road type demanding the strain of extra continuous vigilance for 600 miles in driving as compared with the manifest advantages of a Parkway or valley route. The Highway has bisected the entire Shenandoah National Park, already handicapped by its extremely narrow width. It has left no wilderness areas beyond the sound of travel."

[116] ATC "Resolution," Archives, Shenandoah National Park Headquarters, Luray, VA.

[117] Myron Avery, "Status of the Skyline Drive," *PATC Bulletin* (January 1936) 16-17.

[118] See Lambert, "Administrative History, " 169-171.

[119] See Simmons, "The Creation of the Shenandoah," 11-12.

[120] See Walter J. Showalter, forward to William E. Carson, *Conservation and Development in Virginia: Outline of the Work of the Virginia State Commission on Conservation and Development* (Richmond, VA: Division of Purchase and Printing, 1934) 3.

[121] Showalter, forward to Carson 3; also Foresta.

[122] The National Park Service inventoried the Shenandoah area's natural resources in 1927. See Cammerer, *Report to the Secretary.*

[123] Perdue and Martin-Perdue, "'To Build a Wall Around These Mountains,'" 48 ff. The Perdues cited a 1916-17 study by British folklorist Cecil Sharp. Sharp viewed the culture of the southern Appalachians as unique and susceptible to the loss of its uniqueness through industrialization and creeping progress of the dominant "civilization."

[124] The SANPC also recommended a park in the Great Smokies.

[125] Simmons, "The Creation of the Shenandoah" 18.

[126] Simmons, "The Creation of the Shenandoah" 25.

[127] W.J. Showalter, "Park Will Bring Prosperity to City," Washington Post (April 19, 1926) clipping in General Files; Central Classified Files 1907-1942, 1907-1932, File 0-32; NPS, RG 79; NACP. Showalter touted his ability to estimate the benefits of the park development. "Studying the business significance of the park to Washington, I am persuaded, in the light of my knowledge of urban growth in other cities, my observation of the development of the National Capital and my acquaintance with what the park-bound tourist tide has meant to California and Colorado, that I err on the side of conservation in the following itemization of what the establishment of this great national park on the very horizon of the Washington Monument will mean to the various interests in Washington."

[128] Showalter, "Park Will Bring Prosperity."

[129] See for instance, C.E. Starweather to Colonel Glenn Smith, December 3, 1926, General Files; Central Classified Files 1907-1942, 1907-1932, File 0-32 ; NPS, RG 79; NACP, L. Ferdinand Zerkel to Stephen T. Mather, April 29, 1926, National Parks: Shenandoah; Central Classified Files 1907-1942, 1907-1932, File 0.32; Records of the National Park Service, Record Group 79; NACP, a map showing the route of the proposed Washington Boulevard, undated (1926–1930) Harry F. Byrd Executive Papers, 1926–1930, Virginia State Library, Richmond, VA, and Hugh Naylor to Horace Albright, September 28, 1932, Central Classified Files 1907-1942, 1907-1932, File 631.1-901; Records of the National Park Service, Record Group 79; NACP.

[130] See L. Ferdinand Zerkel, "Shenandoah: The East's First Great Park," clipping attached to a letter from Zerkel to Colonel Glenn S. Smith, May 3, 1926, General Files; Central Classified Files 1907-1942, 1907-1932, File 0-32; NPS, RG 79; NACP.

[131] L. Ferdinand Zerkel to Roy Lyman Sexton, February 14, 1931, Zerkel Papers, Archives, Shenandoah National Park Headquarters, Luray, VA.

[132] Showalter, "Park Will Bring Prosperity."

[133] Zerkel to Sexton, February 14, 1931, Zerkel Papers.

[134] Shenandoah National Park Association Incorporated, Shenandoah: A National Park Near the Nation's Capital (Richmond, VA: The William Byrd P, Incorporated, nd) 4.

[135] Showalter, forward to Carson, 4.

[136] From the November 27, 1927 minutes, Vol. 1, 1926-1927, Programs of Meetings of the State Conservation and Development Commission of Virginia, Record Group 9/F/9/1/1. Virginia State Library, Richmond, VA.

[137] Carson, *Conservation and Development in Virginia* 7.

[138] Carson, *Conservation and Development in Virginia* 7.

[139] Showalter, forward to Carson, *Conservation and Development in Virginia* 4.

[140] See John F. Horan Jr., "Will Carson and the Virginia Conservation Commission," *The Virginia Magazine of History and Biography* 92 (1984): 413. Horan's work provided a good biography of Carson.

[141] See, Hugh E. Naylor to Horace M. Albright, September 28, 1932, and Charles E. Peterson, Memorandum to Horace M. Albright, September, 28, 1932, Central Classified Files 1907-1942, 1907-1932, File 631.1-901; Records of the National Park Service, Record Group 79; NACP. Naylor was president of the John Marshall Highway Association. See also an editorial clipped out of the *Waynesboro News-Virginian* July 23, 1931, Central Classified Files 1907-1942, 1907-1932, File 630; Records of the National Park Service, Record Group 79; NACP, for the Waynesboro issues.

[142] The tourist economy had seemingly supplanted the industrial economy in Waynesboro. The tourist decline since the mid-1970s has had a greater impact.

Chapter 6

The Green Mountain Parkway

Visionaries and Boosters – II

> Vermont has before her a momentous issue. She may lose the larger part of her legitimate share of the Federal public works appropriation, on which nevertheless she must pay the carrying charges. Or she may secure it for the creation of the only project of magnitude suited to her conditions, with which for all time to bring spiritual and material blessings to her own citizens and those of the country at large.[1]
>
> *William J. Wilgus, August 1933.*

1. INTRODUCTION

The proposed Green Mountain Parkway emerged from conflicting visions of regional development. Plans for the parkway, along the ridge and flanklines of the entire length of Vermont's Green Mountains, were voted down in a statewide referendum in 1936. The rejection of the project attested to the public nature of the planning process, the broader issues involved in the development of such large-scale public infrastructure, and the political climate prevalent in Vermont. The debate over these issues, however, demonstrated quite publicly the tenets of the two groups vying for the development of resources in the region. Further, the history of the proposed Green Mountain Parkway shows that the National Park Service had truly institutionalized the planning and building of parkways in national parks by the mid-1930s. That is, after an inauspicious start to planning the Skyline Drive in 1931, the Park Service, by 1934 and 1935, recognized the value of an exhaustive resource survey prior to the beginning of construction.

2. SETTING UP THE CONFLICT

The major players in the debate over the Green Mountain Parkway were involved in a struggle for the future of Vermont. The rift existed, at least on the surface, between support for economic development that may have brightened the day-to-day hardships brought on by the Depression and a struggle over the conservation of Vermont's Green Mountains. This characterization is symmetrical and is one of the dominant scenarios presented by historians.[2]

Although the proponents of the Parkway certainly elevated economic opportunity as a core principle of their argument, it is not true to argue that only opponents of the Parkway were concerned with the preservation and conservation of Vermont's natural resources.[3] Supporters and opponents of the Parkway alike pushed for the conservation of resources. The groups differed in their methods of conservation and their visions for the future of development in Vermont. Parkway opponents added to their concern over the "slashing open"[4] of the Green Mountains the fear of involvement by the federal government in the state and local affairs of Vermonters.[5] At one point the amount of land slated to be turned over to the federal government amounted to approximately 16 percent of the state.[6] In sum, the groups on each side of the Parkway issue had numerous arguments in defense of their respective positions. At times their arguments overlapped, yet each group's constituency remained pretty consistent throughout the history of the Parkway proposal.

2.1 Boosters – Parkway Supporters

The Green Mountain Parkway supporters consisted of an alliance between the Vermont Chamber of Commerce, the Chamber's Executive Secretary James Paddock Taylor, the civil engineer William J. Wilgus, and the National Park Service. This alliance worked for three years, between 1933 and 1936, in an effort to bring the Parkway to Vermont. Taylor and Wilgus worked in concert to bring about positive public relations for the project, aided by the work of the Park Service's consulting and resident landscape architects. The public relations effort consistently pushed the plan as a way to promote Vermont's economic future.[7]

2.1.1 William J. Wilgus

The idea for the Green Mountain Parkway originated with Colonel William J. Wilgus (1865 - 1949), who had retired to Vermont a few years before proposing the Green Mountain Parkway in 1933. In 1933, he was appointed as chairman of a committee of engineers put together by Vermont's Governor, Stanley C. Wilson. Wilgus proposed the Parkway as a means of securing $10 million, Vermont's share of approximately $3.3 billion in federal appropriations under the National Recovery Act of 1933. Wilgus believed the scale of the Parkway and its inherent conservation and development aspects would warrant the federal expenditures.[8] Prior to embarking on the campaign to win the Green Mountain Parkway, Wilgus had a long career as the Chief Engineer with the New York Central railway. Wilgus developed the plan for and oversaw the construction of Grand Central Terminal and the surrounding urban fabric immediately after the turn of the century.[9] In a discussion of the 1990s renovation of Grand Central, the architectural critic at *The New Yorker*, Paul Goldberger, described Wilgus's major contributions in the development of the Terminal. Much of the success of Grand Central – as a station, real estate destination, and formidable icon of New York City – is traceable back to Wilgus.[10]

During World War I, Wilgus organized much of the railroad transportation for the American Expeditionary Forces in France.[11] It's almost ironic that a man so much a part of creating such a well-known urban space would take that knowledge to Vermont in his retirement only to work up a plan for creating a thoroughly regional transportation infrastructure. For Wilgus, transportation planning remained a permanent vocation both prior to retiring to Vermont and later on during his retirement. In 1945 he published a text entitled *The Role of Transportation in Vermont.*[12]

Before retiring to Vermont in the late 1920s, Wilgus participated in the publication of *The Regional Survey of New York and Its Environs* (RSNY & E). In fact Wilgus' work in New Jersey, as part of the RSNY & E, was unique in that it is one of only a limited number of concepts praised by Lewis Mumford in his critique of the plan proposed in the comprehensive Regional Plan of New York and Its Environs.[13] In a report for the RSNY entitled "Transportation in the New York Region," Wilgus proposed outer and inner beltways consisting of both rail and motorways as well as a unified rapid transit passenger system to serve the transportation needs of the New York Region. Further, he proposed unifying the freight distribution system so that cargo transported by rail, boat, and truck would work in a complementary manner. He also, importantly, proposed that the transportation planning include recreational areas conveniently served by rail, highway, and water carriers.[14] Five years before Wilgus' proposal for the Green Mountain Parkway, he was espousing the doctrine of connecting transportation, recreation, and economic development for the health and well-being of a community.

2.1.2 James P. Taylor

James Paddock Taylor (1872 – 1949), as Secretary of the Vermont Chamber of Commerce from 1912 until his death,[15] worked diligently for the Green Mountain Parkway between 1933 and 1936. He had settled in Vermont in 1908 as an Assistant Principal at the Vermont Academy. He had grown up in New York, attended Colgate University, and done some graduate work at Harvard and Columbia Universities. In 1910, Taylor took part in the founding of the Green Mountain Club. The Club sought to better integrate the Green Mountains into the lives of Vermonters and called for the construction of "trails and roads."[16] At the time of its founding, the Green Mountain Club immediately embarked on an effort to construct the Long Trail along the crest of the Green Mountains from the northern border with Canada to the southern border with Massachusetts. Taylor participated in the planning and conceptualizing of the Long Trail, but was not as active in the actual building of it.[17] Moreover, by the early 1930s he was no longer an active leader in the Club. His support for the Parkway, in any event, went against the Club leadership's vehement opposition to the development of the Green Mountain Parkway.

Proponents of the Green Mountain Parkway who believed the Club's opposition to the Parkway went against its own constitution, of course, would later use the Green Mountain Club's early call for "trails and roads" against it. It's reasonable to argue that the conception of the word "road" in 1910 had a different meaning in 1921, and even more so in 1931. Prior to the widespread use of automobiles, roads and trails were used for foot travelers, as well as those using carriages and horses. Moreover, by 1921, the Green Mountain Club had restated its purposes, making no mention of roads as we know them now. Louis J. Paris wrote the Club's statement of "Purposes and Membership":

> The Green Mountain Club exists to bring the people to the Green Mountains and to make the Green Mountains accessible to the people. It builds and maintains trails, especially the Long Trail and its approaches.[18]

Historian Hal Goldman, in his 1995 *Vermont History* article entitled "James Taylor's Progressive Vision: The Green Mountain Parkway," states that Taylor's interest in the Green Mountains grew out of his early years with the Vermont Academy. During this time, Taylor attempted to conceptualize and promote an effort to incorporate the outdoors into the lives of his students and Vermonters as a whole. For Vermonters to make the best use of the Green Mountains, in Taylor's opinion, meant development of and access to the Mountains, rather than the view of the Mountains as a barrier that split the state in two halves and confined the people to the valleys.[19] In 1912, Taylor went to work as Secretary of the Greater Vermont Association, which later became the Vermont State Chamber of Commerce. Just as William E. Carson seemed to find a natural home with the Virginia Commission on Conservation and Development, Taylor's vision of "conservation and development" of Vermont's resources meshed with the economic boosterism associated with the Chamber of Commerce, and the Vermont Chamber of Commerce emerged as the primary institutional Parkway booster.

2.2 Regionalists – Parkway Opponents

The defeat of a statewide referendum that would have authorized the purchase of the land necessary for the Green Mountain Parkway indicated that a majority of Vermonters were against the project. By a 3 to 2 margin, Vermont rejected the Green Mountain Parkway in a referendum held in the spring of 1936. A number of studies have been conducted by political scientists and political historians that attempt to break down the complicated political dynamics that occurred in Vermont during the 1930s.[20] The citizens of Vermont rejected the proposal for a number of reasons, among them, environmental concerns and fear of federal involvement in local and statewide issues. Yet the most influential reason may have been party politics.[21]

The leadership of the Green Mountain Club and its allies, chief among them anti-New Deal newspapers, may well have made their most rigorous arguments by contextualizing the conflict as one centering on development in the region. That is, the methods employed by the clubs and anti-New Deal newspapers to conserve the Green Mountains and to protect Vermont's cultural uniqueness from metropolitanization associated with planning the Parkway came through in their rhetoric and actions.[22] Both the leadership of the Green Mountain Club and the anti-Parkway newspapers presented their opposition publicly and only in part in terms similar to the arguments made by Benton MacKaye and others opposed to the "slashing open" of the ridgelines. The Parkway proponents, meanwhile, took great pains to answer the criticism posed by these groups in hopes of rallying public opinion to the cause. In the end, however, the public opposition to a "skyline road," whether it existed or not, would not seem to have significantly influenced the outcome of the referendum at all.[23]

2.2.1 Green Mountain Club

The Green Mountain Club, beginning in 1910, built and maintained The Long Trail from the Massachusetts border to the Canadian border. By 1930,

270 miles of trails spanned the distance between Massachusetts and Canada along the ridgeline of the Green Mountains. (The Appalachian Trail follows 95 miles of The Long Trail in the Southern half of Vermont.) Since The Long Trail and the Green Mountain Club predated Benton MacKaye's 1921 "Appalachian Trail" piece, MacKaye called for modeling of the Appalachian Trail on The Long Trail. He wrote:

> Specially good work in trail building has been accomplished by the Appalachian Mountain Club in the White Mountains of New Hampshire and by the Green Mountain Club in Vermont. The latter association has built the "Long Trail" for 210 miles through the Green Mountains – four fifths of the distance from the Massachusetts line to the Canadian. What the Green Mountains are to Vermont the Appalachians are to the eastern United States. What is suggested, therefore, is a "long trail" over the full length of the Appalachian skyline, from the highest peak in the north to the highest peak in the south – from Mt. Washington to Mt. Mitchell.[24]

As a member of the Appalachian Trail Conference, the Green Mountain Club was dedicated to constructing and maintaining its portion of the Appalachian Trail. Moreover, many of the issues of importance to the Potomac Appalachian Trail Club also concerned the Green Mountain Club.[25] While the leadership of the Green Mountain Club consistently opposed the construction of the Green Mountain Parkway, the membership as a whole may not have been as uniformly opposed as the leadership stance may have indicated. In a 1934 survey in which half the approximately 1,000 members participated, 196 members voted in support of the Parkway, while 272 members voted in opposition. For its part the leadership voted 14 to 2 against the Parkway.[26] Ultimately, the Green Mountain Club's loud and public opposition to the Green Mountain Parkway, based primarily on its opposition to traditional development and wilderness principles, came from the leadership, rather than the rank-and-file Vermont membership.

2.2.2 Newspapers

As discussed above, Vermont's two major daily newspapers came down on different sides of the Parkway issue. For the most part, the *Burlington Free Press*, in the northwestern part of the state, supported the project. The *Rutland Herald* consistently opposed the construction of the Parkway. Aside from the influence the papers had on the local citizens in the 1936 Parkway referendum, the supporters, especially the Vermont Chamber of Commerce, consistently responded to the concerns espoused by the *Rutland Herald*. James P. Taylor believed that after a thorough educational campaign, Vermonters would come around to the benefits of the Green Mountain

Parkway. He believed this campaign needed to have the backing of the Vermont newspapers, as well as support from other papers in New England and even New York City. After some negative press from the *New York Herald-Tribune*, Taylor wrote to landscape architect and city planner John Nolen and suggested they produce "some positive material" on the Parkway project.[27] He believed in good press for the Green Mountain Parkway – locally and regionally. For instance, to counter negative press by papers such as the *Rutland Herald*, Taylor widely distributed (under the auspices of the Vermont State Chamber of Commerce) a positive discussion of the Green Mountain Parkway prepared in 1934 by Laurie Davidson Cox, resident landscape architect of the Green Mountain Parkway.[28]

3. PLANNING THE GREEN MOUNTAIN PARKWAY

The planning of the Green Mountain Parkway demonstrated that the National Park Service, the Bureau of Public Roads, and the business interests committed to boosting the Parkway embraced a more open and accessible process than in the days of the "Emergency Relief Funds" of 1930 and 1931. By 1933, the 73rd Congress had passed the National Industrial Recovery Act (NIRA) and a number of other New Deal programs during the first three months of Franklin D. Roosevelt's presidency.[29] The Roosevelt Administration attempted to link broad programs designed to simultaneously stimulate the economy, create employment, and promote conservation. These programs continued, through grants, what the Hoover Administration had begun through loans[30] (although the initial funding for the Skyline Drive had, of course, been a direct federal appropriation). In the context of developing public infrastructure that would make the mountains more accessible for recreation while creating a public works program, the William J. Wilgus proposal for the Green Mountain Parkway was an updated and more public version of the Skyline Drive.

3.1 Vermont's Opportunity – The Wilgus Proposal

In 1933, William J. Wilgus proposed that Vermont work with the federal government to build a scenic parkway along 200 or so miles of the Green Mountains. Wilgus framed his Green Mountain Parkway as the only proposal worthy of the federal appropriations made available through NIRA. His proposal was circulated by the Vermont State Chamber of Commerce, which had put its support behind the Green Mountain Parkway from the start. Despite Vermont's apathy to the New Deal (the state voted against Roosevelt in 1932 and would do so again in 1936), Wilgus intended to capitalize on Vermont's sense of fairness and its expectation that Vermont receive its share of the federal outlay. Wilgus called on Vermont to "bestir herself to secure her quota" of the national government's public works appropriation.[31] He said that the Parkway would cost $10,000,000 to build, $7,000,000 of which would be available through the NIRA. Wilgus suggested that the program would benefit Vermonters through employment opportunities in the short term, as well as an expanded economy through tourism over the long term. All told, he expected between 6,000 and 8,000 Vermonters to gain employment in building the Parkway. His view was that the Green Mountain Parkway would benefit all citizens throughout New England and the East through the establishment of another national park east of the Mississippi.[32]

In his argument, Wilgus also linked the Parkway to two broader national themes. First, he believed that the construction of the Parkway would contribute to the rebuilding of the nation's confidence and morale. Second, he argued that the Green Mountain Parkway would further the efforts to improve conservation and forge recreational infrastructure that were major elements of the New Deal's public works programs. Themes of conservation and recreation intersected: the leisure and recreational outlets provided by the Green Mountain Parkway would be spirit-enhancing and healthy, and would demonstrate a commitment to the progress of community. This concept reflected Wilgus' view of transportation planning as outlined in the RSNY piece written in 1928.[33] The proposed Parkway would also conserve sections of the Green Mountains and prevent their degradation through commercial development.

The Wilgus proposal, published as "Vermont's Opportunity," went on to lay out the Green Mountain Parkway in some detail. He estimated that the Parkway would ultimately encompass 1,000,000 acres, more than 16% of the state.[34] Half of this acreage, claimed Wilgus, was already available in the form of state and national forests as well as other land owned by the state and by Middlebury College. The construction of the Parkway would require a strip of land ranging from 800 to 1,000 feet wide. This strip would be supplemented by the land already in public ownership, as well as by future acquisitions. Wilgus proposed that the Parkway follow a route along the flanks of the mountains, crossing from side to side at gaps between higher peaks, then journeying to the crestline to take advantage of the best views. The "skillful" location of the road and management by the National Park Service would prevent the Parkway from causing harm to The Long Trail. Proliferation of "hot-dog"[35] stands (as Benton MacKaye wrote) and filling stations on top of the Green Mountains would be prevented through the regulation of the limited-access Parkway. He argued that the route he favored would leave the peaks unspoiled and spare The Long Trail, although he allowed that where the two overlapped, the Trail would be relocated.[36] Moreover, he argued, the Parkway would leave the wilderness unspoiled by commercialism, excessive over-development, and roadside blight.

Keenly aware of the Green Mountain Club's early opposition to the Parkway, Wilgus carefully argued that the preservation and conservation needs of Vermont would be met through his proposal. In addition, the state's business community, and thereby the entire citizenry, would benefit from the economic development brought about by conservation and preservation efforts. Wilgus was not prepared to allow the concerns of hikers to prevent the entire state from benefiting from such a project. He addressed this point extensively. With regard to the Green Mountain Club's objection to the road's proximity to The Long Trail, Wilgus furthered his argument for the Green Mountain Parkway by opining that the concerns of the Green Mountain Club would only amount to petty opposition intent on interrupting the pursuit of progress. He even cited the Skyline Drive case and concluded that despite concerns over the construction of the road, the Green Mountain Parkway would provide a multitude of benefits for all Vermonters.[37] While Wilgus may not have completely understood the Skyline Drive case, his account of its development served to bolster the argument for his proposal. The Green Mountain Club's concerns, in his view, were minor and interfered with Vermont's greater good. He concluded his proposal by once again arguing for Vermont to claim her fair share of federal relief funds.[38]

3.2 The Chamber of Commerce Backs Wilgus

Wilgus worked closely with James P. Taylor to push for the construction of the Green Mountain Parkway in Vermont. While Taylor spread the word of the economic development benefits of the program, Wilgus secured funding through his contacts in Washington, DC. Taylor also lined up business organizations and associations intent upon boosting prospects for economic development throughout the state. For instance, by August 1933, the Burlington Lions Club and the Burlington Rotary Club had already signed on with Taylor and Wilgus in support of the Green Mountain Parkway.[39]

Taylor contacted two of the better-known park and parkway planners of the day in an attempt to get support for the Green Mountain Parkway. In response to Taylor's inquiry, John Nolen quickly became an advocate of the Parkway and ultimately was named consulting landscape architect for the National Park Service's survey of the proposed route. In July 1933, Nolen wrote to Taylor with regard to the prospect of the Parkway:

> That's what Vermont needs now, and of the Westchester type. We have been making an intensive study of parkway systems, and are all set to do the best possible work for clients. It's no easy job to plan a parkway properly. Let me know if we can help you before the line gets too long.[40]

Taylor immediately picked up upon this praise for the parkways, especially the "of the Westchester type" part. Even by the early 1930s the Westchester parkways were known for their success as facilitators of economic development, as recreational outlets, and as commuter roads.[41]

Taylor also attempted to enlist the support of Frederick Law Olmsted Jr. Olmsted's negative reply disturbed Taylor, yet did not set back the cause. Olmsted began his letter to Taylor by citing the *Boston Herald*'s recent editorial in opposition to the Green Mountain Parkway. Noting that the editorial writer had some "preconceived" opposition to roads in the wilderness, Olmsted's hostility towards the Green Mountain Parkway was primarily along economic lines. He cited Mulholland Drive in the Santa Monica Mountains of California, which turned out to be a waste of funds because most automobile users chose not to use the route, opting instead for better-maintained roads. Since Mulholland had no direct commercial value, the taxpayers were less likely to pay for its upkeep. "There is little doubt now that the Mulholland road was a wasteful piece of extravagance," he concluded.[42]

Perhaps most revealing about these two exchanges between Taylor and arguably the two most prominent park and parkway designers in the United States was that they came down on opposite sides of the issue. Whereas Nolen viewed the Green Mountain Parkway as a way to duplicate the successes of New York's Westchester parkways in the rural regions in Vermont, Olmsted viewed the project as a waste of money because he felt its distance from population centers and its costliness would outweigh its ability to bring a return on the capital outlay.

3.3 The Reconnaissance Survey

Among numerous appropriations approved during the first two years of Franklin Roosevelt's first term was $50,000 for the "Study of the Green Mountains of Vermont as a possible location for a great National Parkway."[43] The study of the Green Mountains began in April 1934 under the aegis of the National Park Service and the supervision of Resident Landscape Architect Laurie Davidson Cox, with John Nolen designated as Consulting Landscape Architect. After ten months of study, Cox and Nolen submitted their findings to the Governor of Vermont for consideration by the

Legislature. The Reconnaissance Survey, as Cox titled it, brought the National Park Service and the Bureau of Public Roads into the debate.

3.3.1 Federal Involvement

In March 1934, Harold K. Bishop, Chief of the Division of Construction at the Bureau of Public Roads, addressed the Vermont Society of Engineers. In his speech, Bishop compared the proposed Green Mountain Parkway with his experiences on the construction of the Skyline Drive. He also compared the Green Mountain Parkway with the latest project picked up by the Bureau, the start of construction on the Blue Ridge Parkway. Bishop explained the benefits reaped by Virginia through the construction of the Skyline Drive, comparing it to the engineering successes of Switzerland and the rest of Europe. He concluded that although there had been some opposition and concern, "the construction of the Skyline Drive had really done wonders for Virginia."[44]

3.3.2 Laurie Davidson Cox Survey

During the first week of April 1934, the National Park Service and the Bureau of Public Roads conducted a preliminary survey of the Green Mountains and issued a short report, which was received favorably by Secretary of the Interior Harold L. Ickes that same month. The Director of the National Park Service, Arno B. Cammerer, commissioned an in-depth study of the Green Mountains in preparation for the construction of the Green Mountain Parkway. The survey and final report on the Parkway was primarily Cox's work (with some input from Nolen), although Thomas Vint, Chief Landscape Architect of the National Park Service, and George Albrecht, Assistant Landscape Architect with the National Park, supervised the work.[45]

The Green Mountain Parkway survey conducted by Cox in 1934 demonstrated unequivocally that the Park Service's understanding of proper survey procedure had progressed since the allocation of "drought relief" funds for the initial construction of the Skyline Drive. Further, when compared to the short memo produced by Vint as he passed through Washington and the area in the proposed Shenandoah National Park in the spring of 1931, the fact that the Park Service built the Skyline Drive but could not build the Green Mountain Parkway seems ironic. The Green Mountain Parkway survey took more than seven months to complete and exhaustively covered the siting, general scheme, and impact of the proposed development.[46] (Figure 6.1)

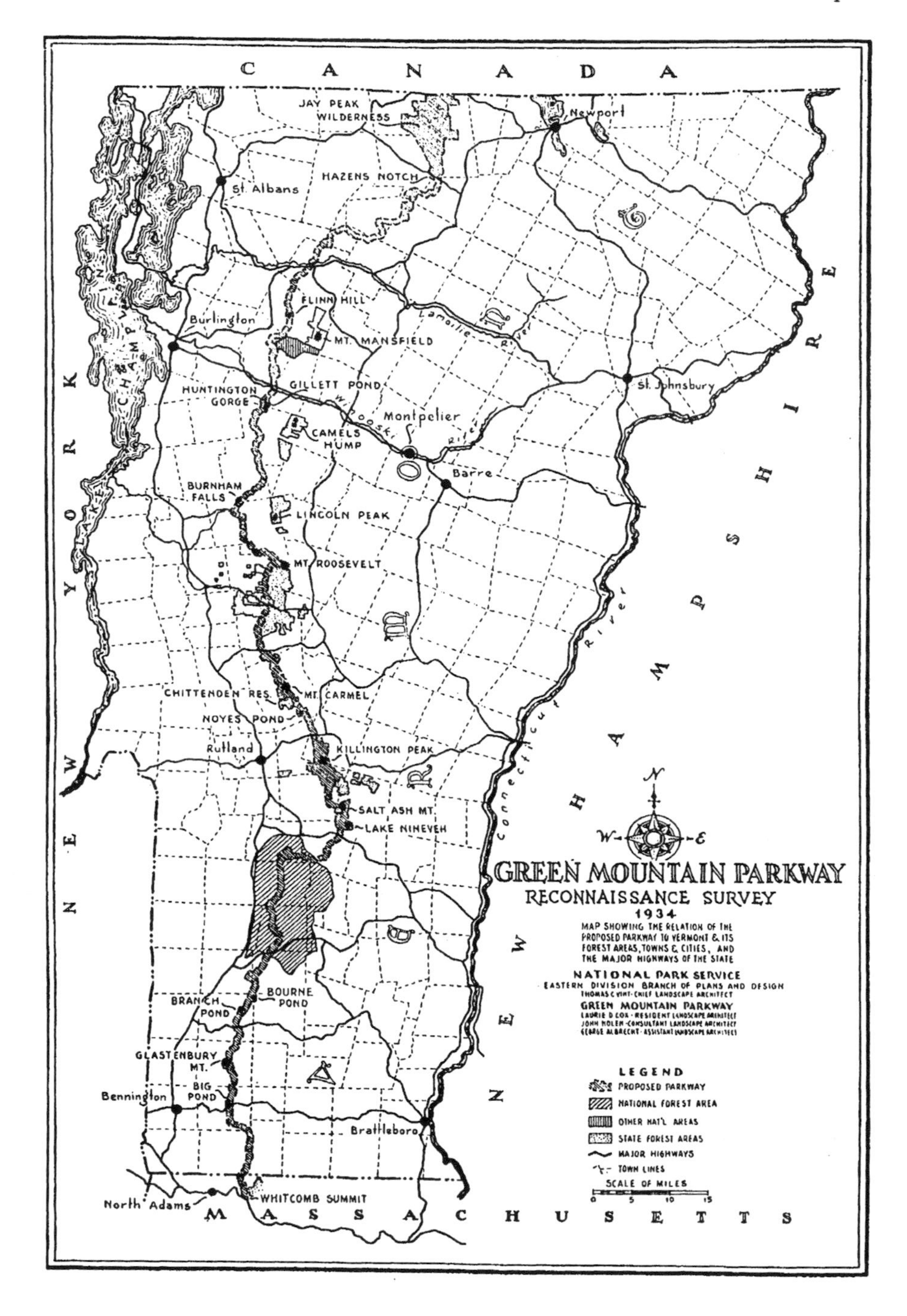

Figure 6.1. Laurie Davidson Cox, "Proposed Green Mountain Parkway," National Park Service, Eastern Division Branch of Plans and Design. 1934

Beginning in April 1934, Laurie Cox worked with engineers of the Bureau of Public Roads to scout out and flag the route of the Green Mountain Parkway. He provided periodic progress reports to Vint and Cammerer. For the most part, Cox progressed quickly because of the availability of topographic maps, favorable weather, and the benefits of Vermont's spring and summer climate. The Parkway was to run the length of Vermont, from the Canadian border in the north to Massachusetts in the south. In Massachusetts it was to connect with the proposed Berkshire Hills Parkway, which in turn would connect with parkways in Connecticut, the Westchester County parkways in New York, through New Jersey, Pennsylvania, and Maryland to Virginia's Skyline Drive, and then to the Blue Ridge Parkway. In addition to the line for the road, Cox proposed the establishment of parallel foot and bridle paths out of sight of the roadway. The footpath would connect with The Long Trail at periodic distances. With regard to The Long Trail, the line chosen by Cox required the relocation of the Trail in some instances, yet after this relocation, the Parkway and The Long Trail were set to intersect only seven times along the 240-mile length of roadway. For 58 miles, or 20% of The Long Trail, the Parkway and the Trail were to lie within one-half mile of each other. Over 22 miles, the Parkway and The Long Trail lay one-half mile to one mile from each other. For the remaining distance, the Parkway and The Long Trail were from one to seven miles apart. Aside from its relationship to the Long Trail, the proposed line of the Parkway was to afford as many different types of views as possible. Cognizant of the view of the landscape from the automobile, Cox wrote of the changing scenery and the nature of the Parkway:

> The Green Mountains of Vermont and the White Mountains of New Hampshire generally excel the other mountain ranges of the East in beauty and wealth of their scenic types [. . .]. Many of these types [of scenic views] will be presented to the traveler from the Parkway road but others (often the finest) will only be secured as a result of short trail trips on foot from the Parkway. All of these types will be available to travelers passing over the foot or bridle paths of the Parkway. Wherever these scenic types are available from the Parkway road they vary considerably in character as to whether they are seen as slowly changing vistas or windshield views, or as rapidly changing side views. [47]

Cox's work reflected a changing attitude toward recreation and the automobile. His recognition of the need for appropriate views from the windshield and the side window was completely new. Further, his discussion of the various views afforded by his route broke with the concept of the Parkway as a "skyline" road. Cox's proposal would afford the traveler the following views over the 240-mile length:

1. Distant views of large lakes.
2. Distant views of mountains or groups of mountains seen across lakes.
3. Views of entire lakes from their shores.
4. Plunging views from above of entire lakes.
5. Views of streams and their immediate banks.
6. Views of streams from their valleys as seen from hillsides above them.
7. Views up or down streams and stream valleys from stream crossings.
8. Views from within stream gorges or chasms.
9. Views of waterfalls.
10. Plunging views from high elevations of definitely composed views of towns and villages.
11. Plunging views from hillsides over wide stretches of wilderness area.
12. Plunging views from hillsides over pastural [*sic*] countryside.
13. Composed views of individual mountains or groups of mountains.
14. Panorama views from high elevations or mountain tops.
15. Views from within of narrow wooded valleys, ravines, passes or notches.
16. Views from within of groves, woodlands, and forest.
17. Views across narrow lakes, streams, ravines or valleys of the vertical faces of steep and precipitous mountain sides.
18. Views from within over open or partially overgrown pastureland and forest clearings.
19. Views from within of wide open valleys with wooded hills as boundaries.[48]

The variety of views offered by the Cox design underscored a difference between the conceptualization of the Skyline Drive and the Green Mountain Parkway. Cox did not design the Green Mountain Parkway as a road along the ridgeline. The Green Mountain Parkway design acknowledged and recognized the Parkway as recreation in and of itself. Indeed, the Cox design reflected many of the principles espoused by Benton MacKaye as he began to accept the parkway as a recreational outlet. MacKaye published his "Flankline vs. Skyline" article in March 1934 in the journal *Appalachia*, describing the benefits of the flankline route as follows:

> The *flankline* tends towards variety of view: its scheme of skirting slopes and winding through the valleys brings out the close-up landscape and

> sets forth the range in all angles; it discloses views both "of" and "from"; the ups and downs display every zone of mountain life from base to timberline; the occasional summit provides a real event.[49]

MacKaye continued by contending that the flankline road would not rip into the experience as the skyline road did. He argued "for scenery and for solitude, both for the motorist and for camper; the flankline principle thus seemed to excel the skyline."[50] Further, in the first footnote accompanying this article, an attack on skyline roads, the editor noted: "[t]he strength of Mr. MacKaye's position, and the especial value of his article, lie in the fact that he is not a mere opponent of such roads but is able to offer an alternative suggestion which would even better serve the purposes of the motorist while not spoiling the mountain for the tramper."[51] Laurie Davidson Cox knew the MacKaye piece, and the Reconnaissance Survey reflected this.

In 1934, the Green Mountain Club sponsored the Appalachian Trail Conference's annual meeting, which Cox attended.[52] Although MacKaye did not attend the conference, he was there in more than spirit. MacKaye had prepared a resolution in opposition to the "skyline or crestline type of highway suggested for the White and Green Mountain Ranges in New England [. . .] and as a substitute [the Appalachian Trail Conference] suggests that whatever highways or parkways are built near the Appalachian Ranges be located along the lower flanks and levels."[53] The Appalachian Trail Conference did not pass MacKaye's resolution. In a letter to MacKaye, Ruth G. Hardy explained the rejection:

> The Vermont members, having recently got assurances that the proposed Green Mountain Drive [Parkway] would be pretty much a "flankline" affair, were frightened that all Federal money would be withdrawn and no road at all would be built. They believed, not without reason, that publicity given to such a resolution, coming from a conference held in Vermont, would be interpreted as opposition to all mountain parkways and further endanger the Federal aid they were seeking.[54]

Rather than risk a vote on the resolution, its supporters withdrew it from consideration.

Hardy had earlier in her letter explained that a number of copies of MacKaye's "Flankline vs. Skyline" article had been available to all members of the Conference. She concluded by describing the conversation she had with Cox regarding the Green Mountain Parkway.

> During the dinner at the Conference, I had sat, quite by accident, with Mr. Cox of the New York State College of Forestry, the landscape architect who has been called in by the Vermont Highway Commission to make the preliminary reconnaissance for the Green Mountain Drive

> [Parkway]. He is definitely a "flankline" advocate; he showed me photographs of spots he is prepared to recommend. In the distance between the Bennington-Brattleboro Road and the Rutland-Woodstock Road, the only section now under consideration, the Long Trail reaches 3,000 feet or higher eighteen times; his proposed highway three times; at most points it is a mile or most distant from the trail. The Vermont people, by all proposing a dozen alternative routes and finally calling in a landscape architect, to curb the power dreams of the raw engineer, have done us all a good service on how this business of "skylines" may be brought down to reasonably good roads.[55]

Clearly, the survey of the proposed Green Mountain Parkway by Laurie Davidson Cox took into account a number of the concerns expressed by Benton MacKaye and the supporters of the ridgeline Appalachian Trail/Long Trail concept. Cox attempted to follow the guidelines set out by MacKaye. In an article in *Landscape Architecture*, Cox spelled out one of the unique features of the Green Mountain Parkway. He explained how for 75 miles, the Parkway would leave the Green Mountains and follow the foothills to the west so as to allow the best view of the most prominent peaks within Vermont – Mt. Mansfield, Camel's Hump, and Lincoln Peak.[56] By even leaving the flank of the mountains, the foothill line not only allowed distant views of the peaks, but also left those areas free of the Parkway, a concept clearly promoted by MacKaye.

In March 1935, Robert Marshall, founder of the Wilderness Society, the era's preeminent wilderness advocate, and an official with the U.S. Forest Service, sent a memo to Secretary of the Interior Harold L. Ickes entitled "The Proposed Green Mountain Parkway." The memo was a follow-up to comments Marshall had made regarding the proposed Parkway before any part of Cox's survey had been completed. Marshall's primary findings based upon the publication of the *Green Mountain Parkway Final Report* elicited an initially positive response:

> The report impresses me as being interesting, imaginative, and enthusiastic. The detailed notes indicate that the proposed route will give the autoist a fine variety of scenery. If the Parkway project could be considered only from the standpoint of scenic highway development, I would certainly endorse it with enthusiasm.[57]

Marshall continued by explaining that the Parkway needed to be considered within the context of The Long Trail. He argued that consideration of the value of the Parkway had to take into account the hikers who wished to travel "by primitive methods." He assessed the impact of the Parkway on The Long Trail and concluded that 50 percent of the 240-mile

Parkway passed within two miles of The Long Trail. Marshall indicated that the central section of the Parkway impacted The Long Trail the least, while the northern section had the greatest impact. After making recommendations for changing the road's line where it most impacted The Long Trail, Marshall noted that while the report was "otherwise splendid," he objected to Cox's use of the term "wilderness" to describe land bisected by the Parkway. Marshall paraphrased Aldo Leopold's definition of "wilderness area" – "extensive areas in which people may travel for several days by their own power without crossing their tracks or encountering the developments of civilization."[58] While Marshall did not completely object to the Green Mountain Parkway proposal, he did put a proprietary mark on the term "wilderness" and ask that it not be used in relationship to the Green Mountain Parkway. Perhaps he was also referring to the use of the term "wilderness" within the context of The Long Trail. In Cox's final report, he pointed out that the existing route of The Long Trail included 29 intersections with roads, 18 of which were major highways.[59] Apparently, just as it may have been difficult to understand that the Green Mountain Parkway was, in fact, a flankline road, it was not easy to comprehend that The Long Trail did not completely remove the hiker from civilization.

Considering the fact that Cox endeavored to plan and design the Green Mountain Parkway off the ridgeline and along the flankline, the criticism of the Green Mountain Parkway design implied a changing of the rules. Cox followed MacKaye's recommendations and created a design that did not "slash open our eastern wilderness." Nearly two years after the rejection of the Green Mountain Parkway, Laurie Cox engaged in a short exchange of letters with Robert Marshall, then the Chief of the Recreation Division with the U.S. Forest Service and a founding member, along with MacKaye and others, of the Wilderness Society in 1935. Cox tried to engage Marshall on the grounds that the Green Mountain Parkway would have preserved some of the "wilderness" promoted by Marshall and the other members of The Wilderness Society. Cox did not have the same foundation for this argument that he had with the "flankline vs. skyline" argument. Marshall recognized this and pointed it out. Wilderness required independence from auto roads, from planning and design, and from civilization. The Green Mountain Parkway could not provide that.[60] In the end, however, Marshall once again cited the Cox report on the Green Mountain Parkway. After again praising it for its vision and design, he noted that The Long Trail and the proposed Parkway line would have passed within a mile of each other for 42 percent of its length and the effect of design on the wilderness precluded the use of the term wilderness.[61] He did not note the intersection of The Long Trail with pre-existing roads, in effect excluding The Long Trail from his own definition of wilderness.

3.4 The Final Report Considered by the Vermont State Legislature

The National Park Service delivered Laurie Davidson Cox's report to the Governor of Vermont in January 1935. The report to the Legislature called for the establishment of a 500- to 1,000-foot right-of-way along the Parkway line staked by the National Park Service and the Bureau of Public Roads. Asked to approve the purchase of 50,000 acres for the right-of-way, the Vermont House of Representatives rejected the bill in March 1935. In a divergent outcome, the Vermont Senate, in the same month, passed similar legislation. In an effort to revive the project, the House of Representatives lowered the minimum acreage to 35,000, yet once again the bill failed to pass. Finally, in the fall of 1935, the legislation came up again. Both the House and the Senate had passed a $500,000 bond issue authorizing funds to purchase the right-of-way, provided that the entire Green Mountain Parkway project pass a state-wide referendum. Vermont held the referendum on March 6, 1936 – Town Meeting Day – and decidedly killed the project by a 12,000-vote margin, out of nearly 74,000 votes cast.[62] Three years after William J. Wilgus formally proposed the Green Mountain Parkway, it had died.

In an attempt to understand the controversy surrounding the Parkway, political historian Frank M. Bryan exhaustively examined the referendum, the relevant politics, and the relative weight of the competing issues. His 1974 analysis concluded that, surprisingly, partisanship played the most prominent role: Vermont during the New Deal was a mostly one-party state.[63]

Bryan seemed ready to assert that the referendum on the Parkway, when framed within the context of economic development versus environmental conservation (Vermont is, after all, named the Green Mountain State), did inform environmental debates 40 years later.[64] He suggested that, given the Green Mountain Parkway case study, the merits of a seemingly nonpartisan environmental argument needed to be cast in a partisan cloak. Of course, it has been 27 years since Bryan's study, and the market-oriented development of the Green Mountains has changed the perspective somewhat. Years of ski resort construction and the complementary market-driven sprawl (albeit better-regulated sprawl than in other states) elicits at least the consideration of an "if only" when reconsidering William Wilgus' call for 1,000,000 acres of land for the Parkway and connected state and national parks and Laurie Davidson Cox's flankline road. [65]

4. MOTIVATIONS AND CONFLICTING VISIONS

Partisanship, which is far from the "boosters versus regionalists" debate, did, however, influence the final referendum on the Green Mountain Parkway on Town Meeting Day in 1936. As an artifact in the history of putting the parkway in the region, the conflict between boosters and regional visionaries, metropolitanists and regionalists did play a significant role. Further, just as Bryan tried to understand how the issues in the referendum may have impacted other more current issues, the divergent views of the regionalists and the metropolitanists informed the contemporary debate. Moreover, the debate influenced the national policy with regard to wilderness, national parks, and parkways, to name three instances. While the Vermont Legislature considered Laurie Davidson Cox's final report, Benton MacKaye, Robert Marshall, Harvey Broome, Harold C. Anderson, and others founded the Wilderness Society, in response to the proliferation of mountaintop roads.[66] In February 1935, just after Laurie Cox had turned over his final report (written under the auspices of the National Park Service and the Secretary of the Interior) to the Governor of Vermont, Secretary of the Interior Harold L. Ickes addressed the Civilian Conservation Corps on the issue of "Wilderness and Skyline Drives." He said:

> I am not in favor of building any more roads in the National Parks than we have to build. I am not in favor of doing anything along the line of so-called improvements that we do not have to do. This is an automobile age, but I do not have a great deal of patience with people whose idea of enjoying nature is dashing along a hard road at fifty or sixty miles an hour. I am not willing that our beautiful areas should be opened up to people who are either too old to walk, as I am, or too lazy to walk, as a great many young people are who ought to be ashamed of themselves. I do not happen to favor the scarring of a wonderful mountain side just so that we can say we have a skyline drive. It sounds poetical, but it may be an atrocity.[67]

4.1 Boosters

The Green Mountain Parkway boosters viewed the Parkway as a mechanism for promoting Vermont's economy through tourism, resource development, and associated commercial prospects brought about by expanded recreational opportunities. The Parkway would also move Vermont forward and provide opportunities for her citizens. Discussion of the Parkway and the newfound strength of regional planning (along the lines of traditional economic development) that went along with the Vermont

State Planning Board and the National Resources Board prompted discussion of Vermont's relationship and importance to New England and the nation as a whole.[68] In a memo regarding state planning, John Nolen, working as a planning consultant to the State of Vermont, wrote:

> [The purpose of state planning is to] promote, through the exercise of foresight, the orderly and beautiful development of Vermont along rational lines, with due regard for health, economy and convenience, and for its commercial, industrial and recreational advancement.[69]

Nolen's memo called for the use of city planning tools throughout the entire state of Vermont. Even after the defeat of the Green Mountain Parkway referendum, the Vermont State Planning Board (perhaps not surprisingly) continued to believe in its own worthiness as a tool of regional planning.[70] Equally important, the Parkway and "rational" state planning meant progress to its leading proponent, James Paddock Taylor.[71]

The Green Mountain Parkway supporters believed the Parkway would promote economic development on a number of fronts. First, the federal funds slated for the construction of the Parkway would be an immediate source of employment and income. During the midst of the Depression, this prospect could only serve to buoy the spirits of Vermont's unemployed. Picking up on this idea, James P. Taylor wrote, "What gives this project a very special interest and meaning to Vermonters at this time is the unemployment relief which it would provide."[72] In promoting the Green Mountain Parkway, the Vermont State Chamber of Commerce seems to have rarely missed an opportunity to tout the economic benefits to all Vermonters. As an example of the lengths to which the Vermont State Chamber of Commerce went, the Editor of the Chamber's publication in which Taylor published his "Vermont's Opportunity" piece included this note:

> This article is of special interest to Vermont truck owners since the parkway construction will require a large number of Vermont trucks and other equipment.[73]

Taylor argued the same thing in a memo to himself written in the first half of 1934. "Does it mean jobs?" he wrote of the Parkway. "Yes [. . .]."[74]

By early 1936, Vermont's allocation of federal funds had increased from $10,000,000 to $18,000,000, which raised the stakes for Vermonters considerably. One argument in favor of the Parkway followed an interesting logic. Since the federal appropriation would be spent on some other state should Vermont reject the project, some supporters argued that rejection would mean the loss of $18,000,000, although Vermont would have to pay its share into the federal treasury regardless. That is, Vermont's federal tax revenue went into the Treasury to pay for the entire federal allocation, and another state would benefit from the tax revenue of Vermont should the state reject the Parkway.[75]

In addition to the short-term influx of jobs and capital, Green Mountain Parkway supporters believed wholeheartedly in the mid-term and long-term economic benefits. Taylor believed Vermont had for too long been isolated from the rest of the Northeast and indeed the country as a whole. During the early 1930s, however, Vermont had come to rely more upon tourism than it did on other, more traditional industries such as dairy and quarrying.[76] Further, economic development from tourism was clean in comparison to other possible industries. William Hazlet Upson, a writer for the *Saturday Evening Post* living in Vermont and a strong Parkway supporter, suggested Vermonters remember the value of the tourist dollar when they considered the Green Mountain Parkway. In a publication promoted by the Vermont State Chamber of Commerce, he wrote:

> As a matter of fact, we don't want to keep the tourists out anyway. Most of them are pretty good people and we want them to come and enjoy our state with us. There are thousands of Vermonters who are glad to supply these summer visitors with board and lodging. In return, the visitors contribute a certain amount of the coin of the realm which comes in very handy in a great many Vermont families [. . .]. In short, no sensible person would want to keep out the tourists – even if it were possible.[77]

Recreation and tourist opportunities required planning and rational decision-making. The alternative, which Hazlett characterized as "idiotic" and "unplanned," had sent Vermont down the road to congestion, commercialization, and overbuilding. Further, unplanned growth would cause Vermont's resources to more closely resemble those of the Adirondacks and the Catskills, where, he wrote, "the natural attractiveness of the countryside has been all but ruined by ill-advised commercial development."[78] Hazlett continued by arguing that contrary to the words of the Green Mountain Parkway opponents, the Parkway would actually alleviate problems associated with roadside attractions. Citing the establishment of other parkways, Hazlett made it known that the federal parkways precluded development typically associated with highways without limited access. "Filling stations, restaurants, and similar conveniences for the traveler would be provided off the main line," Hazlett argued.[79]

Supporters looked outside Vermont for their justifications. Ernest Bancroft, writing in a widely circulated Vermont publication, cited the positive experience associated with the construction of the Skyline Drive in Virginia, and the Blue Ridge Parkway slated to connect the Shenandoah National Park with the Great Smoky Mountains National Park. In *The*

Vermonter, Bancroft, alongside a piece in opposition to the Parkway, wrote of Virginia's "general satisfaction and enthusiasm" for the Skyline Drive, and concluded that Vermont needed to take advantage of the availability of the federal funding opportunity.[80]

Parkway proponents believed tourism and the economic development associated with tourism provided the only viable path to prosperity. Further, the Parkway provided the most rational, well-planned, and consistent mechanism for finding that path. The influx of tourist dollars would, over time, help to revive Vermont's agricultural base and provide capital for the preservation, conservation, and restoration of Vermont's rural communities and historical structures. In short, the Parkway was to be the panacea for all of Vermont's economic aches and pains. Opposition to the Parkway merely mirrored Vermonters' past opposition to things such as the railroad and state highways – contrary opinions for their own sake and out of fear of the unknown. Bancroft ultimately argued for the Parkway's support on all fronts imaginable: economic development, community and statewide importance, conservation, preservation and restoration, the need to preserve additional wilderness areas, recreational opportunities, and the national importance of the Parkway in tandem with other planned eastern parkways.[81] To this day, it is mind-boggling to contemplate the enthusiasm of the proponents and the seemingly endless list of solutions to state problems provided by the Parkway.

Towards the end of 1935, in an effort to bolster the argument for the Green Mountain Parkway, David W. Howe, Business Manager of the *Burlington Free Press*, wrote to a number of newspapers in Central Virginia and the Shenandoah Valley. He asked, "How do the people of your section of Virginia look upon the Shenandoah Parkway development?"[82] Responses from numerous Virginia newspapers, including the *Waynesboro News-Virginian*, the *Richmond Times-Dispatch*, the *Charlottesville Daily Progress*, the *Danville Register*, and the *Winchester Evening Star*, among others, were unanimous in their praise for the Skyline Drive and the prospect of the Blue Ridge Parkway. Nearly every response touched upon the expected tourist dollars generated by the Skyline Drive and the excitement generated by such economic prospects. The responses from the newspapers acknowledged that local citizens appreciated the Skyline Drive and the Shenandoah National Park as a place for recreation for Virginia citizens; however, the greatest pleasure came from the prospect of hosting tourists. The President of the *Waynesboro News-Virginian* wrote:

> Naturally the drive will afford considerable pleasure and opportunity for recreation to our own citizens but primarily it is regarded as an equally desirable opportunity for millions throughout the east.[83]

The *Burlington Free Press*, consistently a proponent of the Green Mountain Parkway, used this anecdotal evidence to bolster its position in the run up to the referendum in March 1936.

Throughout the three-plus-year campaign for the Green Mountain Parkway, supporters looked to the experiences of Westchester County as a model for the proposition in Vermont. Aside from Skyline Drive, which in 1932 had only recently begun construction, the Westchester County Parkways and the Bronx River Parkway were the only examples outside of urban areas. Further, the Westchester examples were very well known. Anticipating the William J. Wilgus proposal of 1933, James P. Taylor organized a trip to Westchester for a series of meetings with, among others, Stanley W. Abbott, Public Information Officer with the Westchester County Parks Commission and later the designer of the Blue Ridge Parkway. Without overstating Taylor's comments, he truly believed that through rational planning and guided development, Vermont could and would experience the suburbanizing influences evident in Westchester. Taylor, in a letter upon his return to Vermont, wrote:

> If more and more Vermonters later from their own initiative and in their own ways study the villages and cities and the highways and parkways of Westchester County, there will seep into the Vermont consciousness more and more the ideas and tastes and desires that we need to inculcate in order to keep things going along the way in which they have been started.[84]

Clearly, for Taylor, progress meant suburbanization along the lines of Westchester County.[85] In a memo "Sent to Weeklies" that echoed his initial letter cited above, Taylor continued along the same lines. Of the Vermonters who attended the meetings, Taylor wrote:

> They were impressed by the fact that every Westchester County community is conscious of being part and parcel of the great regional plan which reaches out 50 miles from New York City in every direction, that such a community has its own local plan, and is probably soon to be a unit in the master plan of a County Planning Commission authorized under a recent New York State Law [. . .]. They heard fair words said about the attractions of Vermont, and came away with many thoughts as to what Vermont can do and must do to keep these attractions unspoiled and perfected, and with the conviction that the more Vermonters study Westchester County, the better for Vermont.[86]

In the end, the boosters of the Green Mountain Parkway viewed the prospect of its construction in a number of ways. Initially, they thought it would have provided the foundation for regional planning and economic development along the lines of Westchester County. As the New Deal took shape and its influence as a tool for economic development grew, the Parkway would have served as the central project for state planning in Vermont, as well as an important foundation for metropolitan planning in New England. Again, that was metropolitan planning along the lines of the Regional Plan of New York and its Environs – metropolitanization. Once the plan proposed by William J. Wilgus had been made public, Parkway boosters turned to the immediate, mid-range, and long-term economic benefits to make their argument. Immediate employment opportunities, new economic development brought about by the tourist trade, and the development brought about by the connection of the Green Mountain Parkway with the rest of the Appalachian Mountains via parkways. And finally, for a state that had historically been landlocked and seemingly trapped by its geography, the Parkway provided a way out, a way of progress that would open Vermont up to new possibilities of progress.

4.2 Regionalists

The regionalists, as represented by the leaders of the Green Mountain Club in Vermont, Benton MacKaye, and the growing advocacy group for the wilderness from afar, provided substantive arguments in the Parkway opponents' efforts to defeat the project in the March 1936 state-wide referendum. Moreover, these groups impacted the method, theory, and practice of putting parkways in the region from the mid-1930s onward. The *ad hoc* group of opponents of the Green Mountain Parkway that had a greater impact upon the rejection of the proposal (anti-New Deal partisans and the *Rutland Herald*, for example) did not share a common regional ideology with Benton MacKaye. These opponents were more than happy to add a mutated form of the regionalist approach to their arguments against the largesse associated with this New Deal program. Landscape Architect Laurie Davidson Cox understood the concerns put forward by MacKaye and Harold C. Anderson with regard to planning the Parkway and the need for wilderness along the ridgeline. So, in effect, the designer, under the auspices of the National Park Service, accepted at least part of the regional vision. The second part of the regional vision – the prospect for social reform – never got off the ground for a number of reasons. First, once again, the Green Mountain Club, like the Potomac Appalachian Trail Club, consisted of primarily middle-class professionals, many of whom lived out of state anyway, and were certainly not interested in the social reform expectations of MacKaye's vision. Second, MacKaye and the anti-ridgeline parkway crowd were so alarmed at the prospect of a parkway running along the Appalachian chain from Canada to the Great Smoky Mountain National Park that they refused to see that the landscape architect had designed a flankline parkway. Finally, the prospect for a social reform-oriented regional plan never got started. Simply preventing the construction of the Parkway was of more concern. Moreover, the middle-class and professional-class membership of the Green Mountain Club had no interest in regional social reform. MacKaye, envisioning great things from the Tennessee Valley Authority (TVA), where he worked from 1934 to 1936, could not even produce the model social-reform planning document he had worked on since the early 1920s. Meanwhile, regional planning in Vermont was effectively left in the hands of consultants such as John Nolen, conceptualized by a traditional economic booster in the person of James P. Taylor, and given its philosophical armature by William J. Wilgus, the engineer of Grand Central Terminal, one of the most quintessentially urban spaces in the United States. Proponents of economic development in Vermont and New England expected to employ traditional, rational urban planning tools throughout the state of Vermont and the New England region. This type of regional planning had neither room for the social reform called for by Benton MacKaye nor expectation of any social reform. Progress meant the metropolitanization of Vermont and New England with bits of parkland set aside for recreation.

In writing his article "The Appalachian Trail: A Project in Regional Planning," Benton MacKaye looked to the founders of the Green Mountain Club and the builders of The Long Trail as the models for his call for the construction of the Appalachian Trail. The middle-class trail builders of The Long Trail, working at times in cooperation with state and national forest personnel,[87] did not hold the same social-reform ideology as Benton MacKaye. Indeed, an examination of James P. Taylor's early work with the Green Mountain Trail reveals a deep-seated desire to open up the mountains and integrate them into the everyday life of Vermonters – a progressive Vermont connected to the economic fortunes of the rest of the country.[88] Arguably, regional social reform as envisioned by MacKaye had long passed from the realm of the possible. MacKaye's experiences with TVA, just beginning in 1934, provided a flicker of hope, yet ultimately proved empty.[89]

When it came down to arguing against the Green Mountain Parkway, MacKaye used as his centerpiece the idea that the Parkway – which he believed would be a skyline road – would destroy the wilderness associated with the Appalachian Trail and The Long Trail. Indeed, the initial discussion of the possibility of the Green Mountain Parkway, courtesy of William J. Wilgus, left a strong impression that another "Skyline Drive" was in the works.[90] For instance, a consultant for Vermont's "Graphic Survey," Philip Shutler, described how he came around to the idea of the Parkway once he had a chance to speak with National Park Service officials and learned about the concepts of design evident in the plan. He explained how he could not support the expense of a "Sky Line Parkway" in the Green Mountains,[91] yet he did come to advocate the idea of the Parkway once he understood that it "would be but little used by people in 'a hurry to go somewhere sitting down,' because this is a leisurely unhurried way."[92] In his article, "Flankline vs. Skyline," MacKaye was seemingly still convinced that the initial Wilgus projection (of a road along the mountaintops) was the fated plan. Published in March 1934 (before Cox had begun to survey the Parkway, but after Wilgus had presented his idea for the Parkway), MacKaye wrote of the proposed skyline parkways:

> The motor road is the logical supplement of the walking trail; a flankline road from Maine to Georgia, properly conceived, would supplement the Appalachian Trail. Such a road should be built [. . .]. I suggest now a New England section of such a road as a substitute for the skyline drives proposed for the Green Mountains and the Presidential Range.[93]

MacKaye spoke out against the Green Mountain Parkway because of his belief that it would be a skyline road. In its stead, he supported an eastern route in New England that would begin an Appalachian parkway, stretching the length of the Appalachian Range. A thorough search of MacKaye's papers did not yield any other discussion or acknowledgement of the Green Mountain Parkway as a flankline road. Finally, by 1934, his efforts in defense of wilderness had taken on a new, more activist meaning. By 1935 he was one of the founders of the Wilderness Society, and his ability to make and argue for compromises such as flankline roads in the region seemed diminished. In the end, however, the evidence is pretty strong that Laurie Davidson Cox designed a "flankline" road in the Green Mountains.

Opposition from the *Rutland Herald* and the anti-Parkway members of the Green Mountain Club seemed to follow along the same lines as those used by MacKaye. The owners of the *Herald* were William H. Field and his son William H. Field Jr. Both were members of the Green Mountain Club and vociferous opponents of the Parkway.[94] Arguments against the Parkway from both the Green Mountain Club and the *Rutland Herald* followed three paths. First, the Parkway would ruin the mountaintops, slash open the wilderness, and displace The Long Trail. This view grew out of Wilgus' initial proposal for the Parkway and, for the most part, endured through the publication of the National Park Service survey by Laurie Davidson Cox. At the beginning of Cox's survey, the leadership of the Green Mountain Club resolved to work with Park Service officials, seeing, as they then thought, that the Parkway was inevitable.

> The trustees wisely decided that it would be useless further to fight the inevitable, but rather [. . .] cooperate with those in charge of the work, in order that it might be carried on with as little damage to the Trail as possible. The route of the parkway has been materially altered from the original Wilgus plan, and it will in general run lower down on the shoulders than it would by the first plan.[95]

This conciliatory sentiment coincided with that of the 1934 Appalachian Trail Conference, which Laurie Davidson Cox attended, and the earliest preliminary reports of the survey work, which were well-received by the National Park Service. The benevolence of the Trustees of the Green Mountain Club towards the Parkway quickly turned sour when, at a September 1934 meeting, the Trustees voted 11 – 0 against any manner of support, no matter what form the Parkway took.[96]

The second path of opposition involved the perceived commercialization of the Green Mountains. Despite information to the contrary, arguments by Green Mountain Club members included the idea that the Parkway would open up the mountains to commercialization, "hot-dog" stands, and "Long Trail Hitch-Hiking." [97] Again, the reality of the plan put forward by Laurie Davidson Cox diminished the validity of this argument, yet its use persisted. The third path of opposition reflected a deep-seated dislike of the New Deal and the perceived "waste" of money on a less-than-important project. The *Rutland Herald* led the way on this course, while the Green Mountain Club Trustees added their voices. As Hal Goldman has argued in "James Taylor's Progressive Vision: The Green Mountain Parkway," the *Rutland Herald* fueled its anti-Parkway stance by consistently referring to Wilgus' initial call for the nearly 1,000,000 acres needed for the completed Parkway. Public ownership of one-sixth of Vermont – in anti-New Deal Vermont no less – did not sit well with the anti-Parkway *Rutland Herald*. Further, it was thought, the federal expenditure should have been allocated for more pressing needs such as flood control and the maintenance of Vermont's roads.[98]

Finally, the Parkway opponents, seeking a way out of the project without the loss of the federal allocation, floated a proposal initiated by the Green Mountain Club Trustees in September 1934. Called the "All-Vermont Plan" and submitted to the Legislature in 1935 by Wallace Fay, the Green Mountain Club Trustees initiated a study after they unanimously rejected any support for the Parkway. A summary of the Trustees' meeting on September 8, 1934, argued that the new plan would "open up attractive locations now inaccessible, and attract a desirable class of people to become summer or permanent residents."[99]

The "All-Vermont Plan" emerged out of this discussion and was sent to the Vermont Legislature. This plan called for an investment in Vermont's existing infrastructure, including family farms and local roads, in the hope of attracting longer-term economic development in the state.[100]

Supporters argued that the "All-Vermont Plan" provided for a more regional solution to the economic needs of Vermont within the context of the offered federal relief aid. Noting the importance of the Green Mountain peaks to Vermont's conception of itself as the "Green Mountain" state, this plan supported economic development in the valleys, a greater series of hiking and bridle trails in the mountains, and funding for structural restorations of historic buildings. The economic development benefits of the "All-Vermont Plan" would significantly outweigh those of the Green Mountain Parkway, according to proponents. Not only would the initial influx of funds benefit more counties than the Green Mountain Parkway, the construction of a series of valley roads connecting towns on either side of the Green Mountains would provide for a more equitable distribution of opportunities. Moreover, these roads would not resemble the "speedway" they believed was slated for construction under the Green Mountain Parkway plan.[101] Finally, the "All-Vermont Plan" would:

> [. . .] leave Vermont in the possession and control of its own citizens as no National Park scheme can. It would avoid dividing the State by a large area of Federally controlled and tax-free National Park land. Planned by Vermonters, it holds to Vermont traditions and ideals. It would not encourage the extension of high-speed, high-pressure methods into the country. It would encourage the healthy and gradual processes of growth that have developed Vermont in the past. It would be thoroughly consistent with the spirit of independence and unwillingness to submit to outside control so characteristic of Vermonters.[102]

Again, this plan would have allowed Vermont to retain some of the federal money allocated to the state, while promoting economic development off the mountains and away from The Long Trail. Although the "All-Vermont Plan" never gained the needed political support, it opened up an interesting issue. A number of the Parkway opponents did support traditional economic development efforts. They viewed the construction of a limited-access highway out of the valleys and away from towns as antithetical to economic development. Discussions of the Parkway as illustrated above – using the word "speedway," for example – led Parkway opponents to believe that visitors would merely speed along the mountain tops between Massachusetts and Canada, bypassing the businesses in the small towns of Vermont.[103] One feature of the opposition was its contradictory assertions; at times there were concerns over "hot-dog" stands in the Green Mountains and at other times there were concerns that the Parkway's limited access would not bring on the needed economic development.

5. CONCLUSION

The analysis of the history of the Green Mountain Parkway and its rejection through public referendum demonstrates that numerous issues influenced the final outcome. The conflicting visions of regional development, on their surface, did not play the major role in the final outcome. Rather, partisan politics on a statewide and national level may have provided the ultimate argument against the Parkway. The conflicting visions did shape the overall debate, however. The referendum result does not detract from the value of the debate over the visions innate in the development of the Parkway idea. In fact, had the referendum turned more explicitly along the lines of the conflicting visions, the overall argument may have demonstrated that the supporters and opponents, as well as the National Park Service officials, had participated in a fine example of open and accessible land-use planning.[104]

The history of the Green Mountain Parkway, when examined within the context of putting parkways in the region, raises a number of important issues. By the middle part of the 1930s the regionalist-versus-boosters-argument had become somewhat fractured. That is, as the Appalachian Trail came nearer to completion, the vision drifted further from the social reform goals first called for by MacKaye. The regional vision evolved into an argument for a connected wilderness trail and for the preservation of the rural environment, not the reformist experiment called for in MacKaye's "Regional Planning and Social Readjustment" memo of the early 1920s.[105]

The regionalists also looked to stave off the metropolitanization of the Green Mountains, but not at the expense of a completed Trail from Georgia to Maine, nor at the expense of the idea that the Trail might at some point later serve as a catalyst for a social-reform movement. The Green Mountain Club, in putting forth its "All-Vermont Plan," called for a vision that would "attract a desirable class of people."[106] Who exactly constituted the implied class of undesirables went unspoken, as did the fact that any such groups' desire for a wilderness experience would mirror that of the Green Mountain Club membership. Viewed in the context of regional planning – of planning in a way that preserved rural values and rural economies and held off metropolitanization – this sort of rhetoric evokes an elite provincialism or even xenophobia. This calls into question the logical extreme of MacKaye's regional vision. Without the social-reform movement intended to complement the Appalachian Trail, did the middle-class professional takeover of the Trail construction lead irreversibly down the path to provincialism? Clearly, MacKaye modeled the trail-building concept on hiking clubs such as the Green Mountain Club. Knowing that it was made up of middle-class professionals content to visit the mountains on weekends and vacation perhaps should have made MacKaye more wary of the danger of the failure of his social-reform movement.

MacKaye's call for the construction of flankline roads instead of skyline roads emerged as a more important issue in the history of the Green Mountain Parkway than with the Skyline Drive. Published in the spring of 1934, his article coincided with the beginning of the survey for the Green Mountain Parkway. As a methodology or blueprint for building regional parkways on or near mountains, skirting wilderness areas, and possibly interfering with the Appalachian Trail, "Flankline vs. Skyline" proposed an attainable solution. The flankline solution was so attainable that, for the most part, Laurie Davidson Cox mirrored many of its ideas. The Green Mountain Parkway went from a skyline road, as proposed by William J. Wilgus, to a flankline road as designed by Cox. Unfortunately for the promoters of the Parkway, if referendum voters knew this, it did not help the cause. The Trustees of the Green Mountain Club knew this, yet they chose to ignore the fact that as a matter of design, the Park Service had called for a Parkway that accommodated The Long Trail. Of course this raises an additional question. Did MacKaye merely propose the "Flankline vs. Skyline" argument as a way to keep the roads not only off the ridgelines, but entirely away from the mountains? Evidence exists that MacKaye knew Laurie Davidson Cox had proposed a flankline road, yet MacKaye did not defend it.[107] It is possible that MacKaye proposed the flankline route as a solution in order to keep roads off the mountain ranges of the East Coast altogether. In fact, as part of the discussion in "Flankline vs. Skyline," MacKaye did propose the Eastern New England route as preferable to the Green Mountain-Berkshire Hills route. Perhaps the Green Mountain flankline route did not adequately overcome the benefits, in MacKaye's mind, of his proposed alternative.

The case of the Green Mountain Parkway demonstrates that broad economic development interests do not always carry the day when it comes to traditional public investment. The most prominent Parkway proponent, James P. Taylor, truly believed that his work with the Vermont State Chamber of Commerce would promote the progress of the state of Vermont. Not only would the Green Mountain Parkway open up the Green Mountains to the citizens of the state, it would open up Vermont to the rest of the world. The benefits of the project would have spanned the state – town-to-town, business-to-business. Taylor carefully studied the situation in Westchester

County, touted that successful development pattern as a model for Vermont, and worked to bring the perceived fruits of metropolitanization to Vermont. To Taylor, Westchester County exuded an aura of progress, and Vermont was sorely in need of an injection of Westchester's planning methods if it were to move forward. Taylor's cohort and initial author of the Parkway plan had participated in the broadest metropolitan survey ever conducted in the United States and was arguably responsible for one of the most recognizably important examples of urban form in North America. Wilgus' work on the Regional Survey of New York and its Environs and Grand Central Terminal undoubtedly influenced his concept for the Green Mountain Parkway. As an engineer concerned with transportation, he valued transportation infrastructure and resources more highly than any other tool of economic development. The health of a community depended upon it. To both Taylor and Wilgus, the health – economic or otherwise – of Vermont depended upon the construction of the Green Mountain Parkway.

Finally, one issue that hardly arose in the Skyline Drive case surfaced a bit more around the Green Mountain Parkway. For the most part, prior to its opening, the Skyline Drive supporters seemed to ignore the idea that the Drive would be a limited-access road. Much of the discussion of the economic benefits of both the Shenandoah National Park and the Skyline Drive glossed over the idea that commercial access would be severely restricted. Economic benefit would come from business establishments sited on private land along the entrance routes to the Park and the Drive. Even today, the small towns along the transverse access roads to the Park have struggled economically, despite, for example, inviting signs such as "Welcome to Stanardsville, Gateway to the Shenandoah National Park." In Vermont, some business interests recognized this issue. The "All-Vermont Plan" and other similarly floated plans called for the construction of roads outside the areas set for control by the National Park Service. That is, why should the business community support a road along which they could not build and grow businesses? While this sentiment was not widespread (had it been, it would have detracted from the anti-Parkway argument that expressed concern over the construction of "hot-dog" stands in the mountains), it did exist and perhaps even demonstrated a crack in the business community's unified support for limited-access, recreational parkways.

---

[1] William J. Wilgus, "Vermont's Opportunity," *Bulletin of the Vermont State Chamber of Commerce* August 8, 1933.

[2] Frank M. Bryan, *Yankee Politics in Rural Vermont* (Hanover: The U P of New England, 1974) 202.

[3] Wilgus, "Vermont's Opportunity."

---

[4] Benton MacKaye characterized the proposed Green Mountain Parkway as "the slashing open of our eastern wilderness belts by skyline roads." See Benton MacKaye, "Re Skyline Drives & the Appalachian Trail," MacKaye Family Papers, Dartmouth College Library, Hanover, NH.

[5] See Hannah Silverstein, "No Parking: Vermont Rejects the Green Mountain Parkway," *Vermont History* 63 (1995): 134-140.

[6] See Wilgus, "Vermont's Opportunity," and Hal Goldman, "James Taylor's Progressive Vision: The Green Mountain Parkway," *Vermont History* 63 (1995): 173. In "Vermont's Opportunity," Wilgus suggested that the Parkway project would include 1,000,000 acres of Vermont's land.

[7] See Wilgus, "Vermont's Opportunity," and James P. Taylor, "Vermont's Opportunity," *Council Headlights* 2:1 (1935): 3; copy in the James P. Taylor Collected Papers, Vermont Historical Society, Montpelier, VT.

[8] Wilgus, "Vermont's Opportunity."

[9] Silverstein 142. Also Goldman 160.

[10] Paul Goldberger, "Now Arriving: The Restoration of Grand Central Terminal is a Triumphant Validation of an Ambitious Urban Idea," *New Yorker* September 28, 1998: 94.

[11] See Earle Williams Newton, "Preface" to William J. Wilgus, *The Role of Transportation in the Development of Vermont* (Montpelier: Vermont Historical Society, 1945) 5.

[12] See Wilgus, *The Role of Transportation.*

[13] Harold M. Lewis, William J. Wilgus, and Daniel M. Turner, "Transportation in the New York Region," *Regional Survey of New York and Its Environs: Transit and Transportation, and a Study of Port and Industrial Areas and Their Relation to Transportation*, Vol. IV (New York: Regional Plan of New York and its Environs, 1928) 161-176; Harold MacLean Lewis, *Regional Survey of New York and Its Environs: Transit and Transportation, and a Study of Port and Industrial Areas and Their Relation to Transportation*, Vol. IV, by Harold M. Lewis, William J. Wilgus, and Daniel M. Turner (New York: Regional Plan of New York and its Environs, 1928) and Lewis Mumford, "The Plan of New York," republished in Carl Sussman, ed., *Planning the Fourth Migration: The Neglected Vision of the Regional Planning Association of America* (Cambridge: MIT P, 1976) 238. Mumford wrote positively of Wilgus' plan for the bay between Sandy Hook and Rockaway Point.

[14] Wilgus, "Transportation in the New York Region" 161.

[15] The Greater Vermont Association hired Taylor in 1912. By 1922 this Association became the Vermont State Chamber of Commerce. Biographical information on Taylor comes primarily from Goldman 160-165, and Laura Waterman and Guy Waterman, *Forest and Crag: A History of Hiking, Trail Blazing, and Adventure in the Northeast Mountains* (Boston: Appalachian Mountain Club, 1989) 353-357.

[16] Goldman 164ff, and Louis J. Paris, "The Green Mountain Club: Its Purposes and Projects," *Vermonter* 16 (1911) 151–171. See also, "Constitution of the Green Mountain Club," Green Mountain Archives, Cowles Papers, Green Mountain Parkway, 1933-1937, Doc. 225, Folder 18, Vermont Historical Society, Montpelier, VT.

[17] Waterman and Waterman, *Forest and Crag* 353–357, and Goldman, "James Taylor's Progressive Vision" 164.

[18] Louis J. Paris, "Purposes and Membership," an excerpt from *The Long Trail Guidebook* (1921) January 15, 1999 <www.greenmountainclub.org>.

[19] Goldman 162-163.

[20] See for instance, Bryan, *Yankee Politics*; Frank M. Bryan and Kenneth Bruno, "Black-Topping the Green Mountains: Socio-Economic and Political Correlates of Ecological Decision Making," *Vermont History* 41 (1973) 224–235, and Richard Munson Judd, *The New Deal in Vermont: Its Impact and Aftermath* (New York and London: Garland Publishing, 1979).

[21] Bryan 233.

[22] Bryan 224.

[23] Bryan and Bruno 234. According to *The Long Trail News*, Vermont members of the GMC voted 117 for the Parkway, 129 against. Out-of-state members voted 79 for, 143 against. See "The Parkway," *The Long Trail News* September 1934, 2.

[24] Benton MacKaye, "An Appalachian Trail: A Project in Regional Planning," *Journal of the American Institute of Architects* 9 (Oct. 1921): 325-330. The Long Trail was extended to the Canadian border by 1930, nine years after MacKaye wrote his article. For The Long Trail today, see The Green Mountain Club website, <www.greenmountainclub.org>.

[25] For instance, the Myron Avery – Harold Anderson split. Avery felt opposing the Skyline Drive would interfere with the completion of the Appalachian Trail. Anderson felt the Skyline Drive interfered too much with the Trail.

[26] "The Parkway," *The Long Trail News* September 1934: 2.

[27] James P. Taylor to John Nolen, September 28, 1934, Taylor Papers, Vermont Historical Society, Montpelier, VT.

[28] See James P. Taylor to John Nolen, September 25, 1934, Doc T12, Taylor Papers, Vermont Historical Society, Montpelier, VT.

[29] Judd 75-85.

[30] Judd 36.

[31] Wilgus, "Vermont's Opportunity" 1.

[32] Wilgus, "Vermont's Opportunity" 1. Of course Acadia National Park (formerly Lafayette National Park) existed, as did the proposals for Shenandoah and the Great Smoky Montains National Park.

[33] See Wilgus, "Transportation in the New York Region" 161; and Wilgus, "Vermont's Opportunity" 2.

[34] Vermont has approximately 5.9 million acres.

[35] Benton MacKaye would use this phrase in "Flankline vs. Skyline," *Appalachia* 20 (1934).

[36] See The Vermont State Chamber of Commerce "Col. William J. Wilgus Explains the Proposed Green Mountain Parkway," *Bulletin of the Vermont State Chamber of Commerce* 28 August 1933, and Wilgus, "Vermont's Opportunity" 3-4.

[37] Wilgus, "Vermont's Opportunity" 4.

[38] Wilgus, "Vermont's Opportunity" 4.

[39] See Goldman 169, and Vermont State Chamber of Commerce, "Col. William J. Wilgus Explains."

[40] John Nolen to James P. Taylor, July 24, 1933, Doc T12, Taylor Papers, Vermont Historical Society, Montpelier, VT. Nolen was referring to the study of parkways and land values he and Henry V. Hubbard were then conducting.

[41] See Stanley W. Abbot, "Ten Years of the Westchester County Park System," *American Planning and Civic Annual*, 5 (1934): 125–126.

[42] Frederick Law Olmsted to James P. Taylor, September 18, 1933, Taylor Papers, Vermont Historical Society, Montpelier, VT.

[43] See Bryan 207.

[44] See Harold K. Bishop, "Park and Parkway Development," Address before the Convention of Vermont Society of Engineers, March 22, 1934, Taylor Papers, Vermont Historical Society, Montpelier, NH.

[45] See Herb Evison, Interview with Dudley C. Bayliss, February 11, 1971, transcript available at the National Park Service Archive, National Park Service Center Archives, Harpers Ferry, WV. Laurie Davidson Cox spent most of his career as Professor of Landscape Architecture at The New York State College of Forestry at Syracuse University (later SUNY College Environmental Science and Forestry). He had a distinguished career as a landscape architect completing designs such as Lincoln Park, Lincoln Park Conservatory, and Griffith Park in Los Angeles, as well as a set of athletic fields in Mamaroneck, NY. He also had an extremely successful career the lacrosse coach at Syracuse University and as the coach of the All-American Lacrosse Team. See Raymond J. Hoyle and Laurie Davidson Cox, eds., *The New York State College of Forestry at Syracuse University: A History of its First Twenty-Five Years, 1911 – 1936* (Syracuse, NY: The New York State College of Forestry, 1936) 56, and McClelland, Linda Flint, *Presenting Nature: The Historic Landscape Design of the National Park Service* (Washington, DC: Government Printing Office, 1994) 34.

[46] See Laurie Davidson Cox, Thomas C. Vint, *et al.. The Green Mountain Parkway Final Report by the Landscape Architects of the National Park Service* (Washington, DC: GPO, 1935).

[47] See Cox, *et al., The Green Mountain Parkway Final Report* 15; also 48ff; also see Cox, *et al., Green Mountain Parkway Reconnaissance Survey* 4-8.

[48] Cox, *et al, The Green Mountain Parkway Final Report* 17-18.

[49] Benton MacKaye, "Flankline vs. Skyline," *Appalachia* 20 (1934): 107. The italics are MacKaye's.

[50] MacKaye, "Flankline vs. Skyline" 104ff.

[51] MacKaye, "Flankline vs. Skyline" 104, footnote 1.

[52] Ruth G. Hardy to Benton MacKaye, July 16, 1934, MacKaye Family Papers, Dartmouth College Library, Hanover, NH. Mrs. Hardy wrote MacKaye of her conversation with Cox at the meeting.

[53] Benton MacKaye, "Expression of Sentiment of the Sixth Appalachian Trail Conference, Regarding Skyline Highways," Long Trail Lodge, Vermont, June 30, 1934, MacKaye Papers, Dartmouth College Library, Hanover, NH. A cleaner version of this memo is called "Re Skyline Drives & The Appalachian Trail," July 14, 1934, MacKaye Papers, Dartmouth College Library, Hanover, NH.

[54] Hardy to MacKaye.

[55] Hardy to MacKaye.

[56] Laurie Davidson Cox, "The Green Mountain Parkway," *Landscape Architecture* 25 (1935): 124.

[57] Robert Marshall, "Memorandum to the Secretary [of the Interior], the Proposed Green Mountain Parkway," Wilderness Society Collection, Denver Public Library, Denver, CO.

[58] Marshall.

[59] See Cox, *et al, Green Mountain Parkway Final Report* 50.

[60] In the final report, for instance, Cox wrote of section in the southern most part of the Parkway – "[the Parkway] will cross through the largest section of wilderness (unbroken by roads) which exists in the state, and will pass through the extensive forest area of the Green Mountain National Forest." See Cox, *et al, Green Mountain Parkway Final Report,* 20. Clearly, a wilderness area with a road through it is no longer wilderness.

---

[61] See Laurie Davidson Cox to Robert Marshall, January 12, 1938, and Robert Marshall to Laurie Davidson Cox, January 20, 1932, Records of the U.S. Forest Service, Record Group 95, Entry 85, U-Recreation Areas (Primitive, Roadways, Wilderness) 1936–1939; NACP.

[62] For the history of legislative actions, see Judd 85ff, Bryan 208-212, Silverstein 150-152, and Goldman 176.

[63] Bryan 231-232.

[64] Bryan published his study in the early 1970s.

[65] Bryan 231-233.

[66] See *The Living Wilderness* 1 (1935).

[67] Harold L. Ickes, "Wilderness and Skyline Drives," *The Living Wilderness* 1 (1935): 12.

[68] The Vermont State Planning Board, the New England Regional Planning Commission, and the National Resources Board followed a process of regional survey and the formulation of a plan to facilitate traditional economic development interests. In contrast, the establishment of TVA (Tennessee Valley Authority) raised the prospect of regional planning ideology associated with the Regional Planning Association of America. Benton MacKaye's initial excitement over TVA and his hiring by the agency in 1934 proved disappointing in the end. MacKaye left after only 26 months with TVA. Daniel Schaffer wrote of MacKaye's ideas, "MacKaye came to TVA with more than the Tennessee Valley's immediate economic and resource problems on his mind; rather, he was enticed by a larger national vision." Of MacKaye's role, Schaffer wrote, "MacKaye is largely forgotten within the agency." See Daniel Schaffer, "Benton MacKaye: The TVA Years," *Planning Perspectives* 5 (1990): 5-21.

[69] John Nolen, "Orientation Statement of the Purpose, Objectives and Organization of State Planning for Vermont," December 29, 1934, Doc. T 9, James P. Taylor Papers, Vermont Historical Society, Montpelier, VT. There is a significant history of state planning and regional planning in New England that needs examination. Although this history intersects with the Green Mountain Parkway, it is outside the scope of this book. Briefly, the history of state planning in Vermont and regional planning in New England came after the introduction of the Green Mountain Parkway and similar New England parkways (the Berkshire Hills Parkway cited by Laurie Davidson Cox, for example). The Green Mountain Parkway predated the call for and organization of the New England Regional Planning Commission, not *vice versa*.

[70] See the New England Regional Planning Commission, *Six for One and One for Six* (Boston: New England Regional Planning Commission, 1937) 53, Vermont Historical Society, Montpelier, VT.

[71] See Wilgus, *The Role of Transportation in the Development of Vermont* 9, Goldman 158-159, and John Nolen, Philip Shutler, Albert La Fleur, and Dana A. Doten, "Graphic Survey: A First Step in State Planning for Vermont: A Report Submitted to the Vermont State Planning Board and National Resources Board" (1935) 12-13.

[72] Taylor, "Vermont's Opportunity" 3.

[73] Taylor, "Vermont's Opportunity" 3.

[74] James P. Taylor, "Memo – Is it Good Business?" undated (Jan.– June, 1934) Taylor Papers, Vermont Historical Society, Montpelier, VT.

[75] See Ernest H. Bancroft, "Why People Should Favor Green Mountain Parkway," *The Vermonter* 41 (Jan. – Feb. 1936): 5-8.

[76] See Silverstein 135.

[77] William Hazlett Upson, "The Green Mountain Parkway," *Informational Bulletin on State Problems* 4 (Burlington: Vermont State Chamber of Commerce, 1934) 1.
[78] Upson 2.
[79] Upson 3.
[80] Bancroft 6.
[81] Bancroft 5-8.
[82] See E.W. Opie to David W. Howe, January 1, 1936, Doc. T3, James P. Taylor Papers, Vermont Historical Society, Montpelier, VT.
[83] Louis Spilman to David W. Howe, January 1, 1936, Doc. T3, James P. Taylor Papers, Vermont Historical Society, Montpelier, VT.
[84] James P. Taylor to W. H. Beardsley, June 21, 1932, Doc. T13, James P. Taylor Papers, Vermont Historical Society, Montpelier, VT.
[85] For an analysis of the roots of James P. Taylor's understanding of progress, see Goldman.
[86] James P. Taylor, Memo "Sent to Weeklies," June 27, 1932, Doc T13, James P. Taylor Papers, Vermont Historical Society, Montpelier, VT. Taylor was obviously using his knowledge of the Regional Plan of New York and its Environs and the Regional Survey (which William J. Wilgus contributed) to embellish his understanding of the planning situation in Westchester County.
[87] See Waterman and Waterman 353ff.
[88] See Paris 168, and Goldman 164ff.
[89] Schaffer 5-21.
[90] Without making too great an issue of the validity of their arguments, I believe a number of historians who have looked at the Green Mountain Parkway believe it was to be a "skyline" road – other historians have viewed the controversy from different perspectives. Laurie Davidson Cox's final report contradicts this, as does the conversation at the Appalachian Trail Conference reported to Benton MacKaye in a letter by his friend Ruth G. Hardy.
[91] Philip Shutler to F. W. Shephardson, January 28, 1936, Doc. T13, James P. Taylor Papers, Vermont Historical Society, Montpelier, VT. "[. . .] there was neither engineering nor economic justification for the expense of climbing a peak from the south only to descend on the north, and then immediately start the ascent."
[92] Shutler to Shephardson, January 28, 1936, Doc. T13, Taylor Papers, Vermont Historical Society, Montpelier, VT.
[93] MacKaye, "Flankline vs. Skyline," 108. MacKaye wrote this article before Laurie Davidson Cox had even begun to survey the line for the proposed parkway.
[94] See "William H. Field, Obituary," *The Long Trail News* (April 1935): 3. See also Goldman 172.
[95] "Special Meeting of the Trustees," *The Long Trail News* (June 1934): 2. Vermont Historical Society, Montpelier, VT.
[96] "Trustee's Meeting," *The Long Trail News* (September 1934): 1. Vermont Historical Society, Montpelier, VT.
[97] "Some More Opinions," *The Long Trail News* (November 1933): 1. Vermont Historical Society, Montpelier, VT.
[98] Goldman 171-173. In 1933, 40,000 acres in the proposed parkway area were in the National Forest.
[99] "Trustee's Meeting" 1.
[100] Silverstein 149.

---

[101] See the Journal of the House of the State of Vermont, Biennial Session, 1935, "The All-Vermont Plan" 232-234. The plan consistently referred to the Green Mountain Parkway as a "speedway,"

[102] "The All Vermont Plan" 232-234.

[103] See "No Green Mountain Hot-Dogs," *The Literary Digest* 121 (1936): 9, and Arthur Wallace Peach, "Proposed Parkway a Threat to the State's Well Being," *The Vermonter* 41 (1936): 8-13.

[104] An alternate view, perhaps beyond the scope of this book and within the realm of political science, posits the influence of partisanship on the nature of the debate between the regionalists and the metropolitanists.

[105] See Benton MacKaye, "Memorandum: Social Planning and Social Readjustment" MacKaye Family Papers, Dartmouth College Library, Hanover, NH.

[106] "Trustee's Meeting" 1.

[107] See Hardy to MacKaye, July 16, 1934, MacKaye Family Papers.

Chapter 7

Conclusion

> Instead of creating the Regional City, the forces that automatically pumped highways and motor cars and real estate developments into the open country have produced the formless urban exudation. Those who are using verbal magic to turn this conglomeration into an organic entity are only fooling themselves. To call the resulting mass Megalopolis, or to suggest that the change in spatial scale, with swift transportation, in itself is sufficient to produce a new and better urban form, is to overlook the complex nature of the city.[1]
>
> *Lewis Mumford, 1961.*

1. THE STATE OF THE HISTORICAL DEBATE

Benton MacKaye and Lewis Mumford were correct in fearing urbanization of the region. Metropolitanism has engulfed much of the region in both Vermont and Virginia, as well as in many other places across the United States. The economic realities that drove the metropolitanists to push for their conception of the parkway in the region prevailed throughout the last 65 years of the twentieth century. Robert Moses' displacement-happy endeavors that created the Cross Bronx Expressway and the Northern and Southern Parkways,[2] sprawl-swept southern and sunbelt metropolitan areas, the continuation of the interstate highway frenzy, and even urban renewal, all have roots in the metropolitanist reality eloquently defended by Thomas Adams in his "A Communication in Defense of the Regional Plan."[3] The conflict between the regionalists and the metropolitanists not only shaped the regional parkway, but it shaped our contemporary urban, suburban, and regional landscapes.

The regional parkway evolved during an era defined by the impending decentralization of American cities. The regionalists subscribed to the concept of the Fourth Migration as articulated by Lewis Mumford. The metropolitanists perhaps did not follow an identifiable theory that described the decentralization. The developers of the Bronx and Westchester Parkways, and the effort of the Russell Sage Foundation to promote the Regional Plan of New York and Its Environs, demonstrate that the metropolitanists understood the connection between declining congested central cities and the forces contributing to decentralization. Moreover, the rising automobile culture and consumer culture thrived on the efforts of the metropolitanists to facilitate economic development.

Anticipating the decentralization of the American city, the regionalists recognized two possible paths. The first evoked images of an urbanized region choked and congested by untrammeled, dehumanizing industrialization. Clarence Stein described the creeping urbanization with the term "Dinosaur Cities."[4] These cities, metropolitan areas really, required constant fuel for development and depended upon economic growth and population increases for perceived viability. Cities that did not pursue this strategy lost out. As Frederick Ackerman wrote, "[a] community with a stable population is now referred to as a 'dead one.'"[5] The alternative path to megalopolis and dinosaur cities involved regional planning – a recentralization – that at once revitalized congested urban areas through decentralization and also repopulated the region on a human scale. The regionalists believed that following this path merely required tweaking market forces and educating the population. It seemed as natural for an industrial worker to desire periodic escape to the ridgeline of the Appalachian Mountains as it would for the same worker to desire permanent relocation to a garden city outside the congestion of the dinosaur city. Reform required educational examples such as the Appalachian Trail, Patrick Geddes' survey concept, garden cities, open space preservation, and the efficient cultivation of our natural resources.

The metropolitanists did not necessarily follow, however, nor did they intend to follow, the congestion-filled metropolitan path disdained by the regionalists. This group, which included planning theorists and practitioners, landscape architects, business boosters, and automobile associations, believed that orderly decentralization of urban congestion required the rational application of city planning tools. That is, the metropolitanists used the urban policies that led to the city beautiful and the city efficient in the region, without acknowledging that these tools were in part responsible for the congestion in the first place. The metropolitanists planned and worked to order regional growth, suggesting that some "restriction[s] on the rights of property, but no further than it is reasonable to expect public opinion to go, or government to authorize in the future,"[6] would help to order the recentralization when development "is injurious to public well being."[7] Aside from obvious instances of civic transgressions, the metropolitanists answered to and promoted economic development interests. The metropolitanists pursued their agenda through rational planning strategies, survey, the facilitation of capital accumulation, and the promotion of local economic development interests.

The regional parkway presented a unique opportunity to both the regionalists and the metropolitanists. The regionalists believed the regional parkway was the leading edge of the metropolitanization of the region. Benton MacKaye had associated creeping metropolitanization with a rising flood or tide, invoking images of a universal, all-encompassing fluid that seeped into the uniqueness of the region. This liquid tended to drown out the small town and the civilizing nature of regional difference. Moreover, MacKaye called upon regional planning to act as a dam against a rising sameness. He imagined the dam protecting the small town and the foot traveler.

To the metropolitanists, the parkway provided for a convenient and rational infrastructure that could, at once, facilitate decentralization and economic development, and provide recreational outlets for the metropolitan citizen. Further, while the regional visionaries valued the humanizing attributes of the small town and the foot traveler, the metropolitanists viewed the region as the realm of the automobile and as the place where traditional development spurred on decentralized economic opportunity regardless of place.

The case studies show how the two conflicting ideologies shaped the process of putting the parkway in the region. While the Green Mountain Parkway was never developed, the region bisected by the Skyline Drive only thrives economically insofar as it is linked to the economies of Northern Virginia and Washington, D.C. to the north, and Charlottesville, Richmond, and Central Virginia to the south and east. Metropolitan growth may not physically rest at the edge of the Skyline Drive, but its influence certainly does. The regional uniqueness of the culture of the mountain folks and those of the agrarian foothills of the Piedmont in Virginia is lost. The mountain culture is a topic studied by local anthropologists and genealogists, while the historical Piedmont farmstead is under consistent assault from speculative developers intent on converting the next corner's hayfield into the market-tested Super Wal-Mart. The history of the Skyline Drive demonstrates that Drive proponents put their own economic interests at the fore, sacrificing the reality of the small town and its unique culture at the altar of consumption and the automobile. Skyline Drive and Shenandoah National Park proponents went so far as to construct a wilderness vision of the Blue Ridge Mountains, despite the existence of human scale elements of regional culture, so that, in the end, the unique places could be subsumed by the sameness of metropolitan refinement.

The Green Mountain Parkway case is an inverted mirror image of the Skyline Drive. The rejection of the Green Mountain Parkway by the citizens of Vermont in 1936 only started the battle over Vermont's regional resources. Through the 1960s and 1970s, the ski industry developed the Green Mountains and attracted many times the number of tourists projected for the Green Mountain Parkway. While this work does not address the extent to which the Parkway would have displaced families and individuals for its development, it is interesting that market-rationalized purchases of land do not carry the same meaning as "Federally controlled and tax-free National Park land."[8] William Wilgus' Green Mountain Parkway did not bring the universal culture of the metropolis to Vermont. Instead, market forces and the uniqueness of the geography in Vermont opened the floodgates and allowed a form of urbanization to creep into the regional culture. It is ironic that the design of the Green Mountain Parkway satisfied many of the articulated design concerns of Benton MacKaye, yet, because the plan did not earn acceptance by the public, the Parkway's mitigating effect on market forces along the ridgeline never materialized. No doubt "hot-dog" stands and ski lifts share the same topographic line as The Long Trail on many a peak in the Green Mountains today.

In *All That is Solid Melts Into Air*, Marshall Berman recounted his deeply felt antagonism toward Robert Moses and the Cross Bronx Expressway. Berman's analysis of the impacts and implications of modernity, especially the modernity of automobile infrastructure, helps flesh out an understanding of the Green Mountain Parkway and the Skyline Drive. According to Berman, the Cross Bronx Expressway not only destroyed the traditional and the uniquely cultural institutions of pre-modern Bronx community life, it also replaced modern icons with other, more modern icons.[9] Clearly, the Skyline Drive replaced the pre-modern culture and traditions of the mountain folk and the regional economies of Central Virginia and the Shenandoah Valley. Moreover, just as Berman ruefully looked over the Cross Bronx Expressway and had difficulty separating the road from the very meaning of the physical place in which he grew up, so too did the mountain folk displaced by the Drive. Former landmarks known throughout the mountain communities are now replaced by National Park Service interventions such as picnic grounds, monuments, and old CCC camps.[10]

The Green Mountain Parkway again presents the inverted mirror image of the Skyline Drive. Filtered through Berman's analysis, by rejecting the

Green Mountain Parkway referendum, the citizens of Vermont rejected the modernity of the urban boulevard in 1936. This rejection of modernism may merely have laid the foundation for the market to deliver the modernity of the ski industry thirty years later. Moreover, the rationality of the market has led to year-round operation of the ski resorts, using whatever modern consumer attraction sustains the industry – bicycles, extreme sports, and others. Benton MacKaye's discussion of the young yahoos in "Flankline vs. Skyline" becomes clearer each passing year.

The two ideologies poised to shape the parkway in the region intended to guide the recentralization of populations on the move because of modernity's impact on urbanization. The metropolitanists, as described by Mumford and the RPAA, also accepted the flow of populations back into the large urban areas, as the cause of urban congestion. Technological advances in industry and transportation, as well as the cultural attractions of the central city were all products of modernity. Further, the era of decentralization, again facilitated by transportation advances, had similarly modern roots. It would be easy to describe the theories associated with the RPAA as anti-modern, Luddite, or reactionary. This would be wrong. Just as the metropolitanists tried to rationalize the decentralization of urban congestion, the regionalists wanted to use modern technology – Giant Power, technological advances in transportation, efficient distribution of natural resources and products to reinvigorate the region. Since the small town and the cultural resources of the region were inherently American, as Mumford wrote in *The Golden Day*, the introduction of the tools of modernity were to be used to promote the regional culture, not to destroy it. It was the metropolitanists' conception of the regional parkway that contributed to the destruction of the regional culture. The Appalachian Trail and the introduction of technological infrastructure were intended to bolster the regional economy, not destroy the regional culture.

Benton MacKaye believed his Appalachian Trail project carried on the tradition he had picked up during his childhood when he and his brother explored the woods near their home and characterized these exercises as "Expedition Nining." This process of exploration recalled Henry David Thoreau's mid-nineteenth century discourse on the New England landscape in *The Maine Woods*, *A Week on the Concord and Merrimack Rivers*, and *Walden*. While Thoreau may have been synthesizing an early reaction to modernism, even lamenting the destructiveness associated with populations creeping into the wilderness, he did praise the value of the wilderness in that era of flux.[11] MacKaye's Appalachian Trail and the desire of the RPAA to halt creeping metropolitanization didn't constitute an anti-modern, anti-technological, anti-civilization movement. Regional planning intended to re-civilize the region in a manner appropriate to the region. The regional parkway could not aid that re-civilizing process because of its inherent metropolitan foundation.

2. TIME FOR A NEW REGIONAL VISION

The 1996 update of the Regional Plan of New York & Its Environs, produced by the Regional Plan Association (RPA) and titled *Region at Risk: The Third Regional Plan for the New York-New Jersey-Connecticut Metropolitan Area*, immediately invokes the legacy of the split between Thomas Adams and Lewis Mumford. In the introduction, the plan's authors, Robert D. Yaro and Tony Hiss, proclaim victory for Adams and relegate Mumford's arguments to the graveyard of failed philosophies.[12] Yaro and Hiss go on to praise the efficient growth and continued increase in quality of life within the region as "master builders" such as Robert Moses, Austin Tobin, and John D. Rockefeller carried out segments of the 1929 plan.[13]

By rejecting Mumford's ideas out of hand through the comparison of his ideas with the downfall of centralized planning as envisioned by totalitarian states, the authors of *The Third Regional Plan* did not have to acknowledge the downside of metropolitan planning at all. The metropolitanists had won the debate over the region and the idea of regional difference and regional uniqueness, and the landscape of the small town had passed from the debate. In an ironic twist that demonstrates that political ideologies remain hard and fast, to the detriment of a thorough, invigorating debate over regional planning, conservatives criticized *The Third Regional Plan.* A columnist in *The New York Post* wrote, "Unembarrassed by the disrepute into which central planning has fallen around the world, the RPA continues to churn out plans for the Tri-State region."[14] In a poisoned political climate without a platform for conducting a reasonable debate over regional planning, where should a regional plan effort turn?

The hope of the region lies in the grassroots. The sustainable communities initiatives proposed by small towns, cities, and regions throughout the United States use the rhetoric of the Regional Planning Association of America, yet also create supportive constituencies beginning with the earliest survey efforts. It is a lesson learned from this conflict over planning in the region.

---

[1] Lewis Mumford, *The City in History* (San Diego, New York, London: Harcourt Brace Jovanovich, Publishers, 1961) 505.

[2] See Robert A. Caro, *The Power Broker: Robert Moses and the Fall of New York* (New York: Vintage Books, 1974) 299–303, and Robert A. Caro, "The City Shaper," *New Yorker*, January 5, 1998: 38-55.

[3] See Thomas Adams, "A Communication in Defense of the Regional Plan," *The New Republic* 71 (1932): 207-210.

[4] See Clarence S. Stein, "Dinosaur Cities," *The Survey* 54 (1925): 135-138.

[5] Frederick L. Ackerman, "Our Stake in Congestion," *The Survey* 54 (1925): 141-142.

[6] Adams 208.

[7] Adams 208.

[8] See Journal of the House of Representatives of the State of Vermont, Biennial Session, 1935, "An All-Vermont Plan" (Published by Authority, Montpelier, VT, 1935) 232.

[9] Marshall Berman, *All That is Solid Melts Into Air* (1982; New York: Penguin Books, 1988) 295.

[10] Virginia Taylor, personal interview by the author, 25 February 1994.

[11] See Henry David Thoreau, *The Main Woods* (1864, New York: Book of the Month Club, 1996).

[12] Regional Plan Association, Robert A. Yaro, and Tony Hiss, *A Region at Risk: The Third Regional Plan for the New York-New Jersey-Connecticut Metropolitan Area* (Washington, D.C: Island P, 1996) 1.

[13] Regional Plan Association 2.

[14] Irwin Steltzer, "Don't Believe the Doomsayers," *The New York Post* 21 March 1996.

Sources

Published Sources

Abbott, Stanley W. "Shenandoah-Great Smoky Mountains National Parkway." *American Planning and Civic Annual* 6 (1935): 31-33.

___. "Ten Years of the Westchester County Park System." *American Planning and Civic Annual* 5 (1934): 125-126.

Adams, Thomas. "A Communication in Defense of the Regional Plan." *The New Republic* 71 (1932): 207-210.

___. "Planning for Civic Betterment in Town and Country." *The American City* 15 (July 1916): 47-51.

___. "Regional Highways and Parkways in Relation to Regional Parks." *The Proceedings of the Nineteenth National Conference on City Planning.* Philadelphia: Wm. F. Fell, 1927.

Allen, Harold. "The Shenandoah Scenery." *American Civic Annual* 11 (1930): 15-18.

Allen, Harold, and L.F. Schmeckebier. "Shenandoah National Park: The Skyline Drive and the Appalachian Trail." *Appalachia* 22 (1936): 75-83.

Anderson, Harold C. "The Good Old Days." *Bulletin: Potomac Appalachian Trail Club* 21 (1952): 88-93.

___. "Primitive Trails and Super-Trails." *The Living Wilderness* 1.1 (1935): 8.

___. "What Price Skyline Drives?" *Appalachia* 21 (1935): 408-415.

Annese, Domenico. "The Impact of Parkways on Development in Westchester County, New York City, and the New York Metropolitan Region." *Parkways: Past, Present, and Future.* Boone, NC: Appalachian Consortium P, 1987. 117-121.

Asheville Citzen, 26 November, 1931.

"Auto Trail on Top of Blue Ridge if Picked for Park." *Harrisonburg Daily News-Record* October 25, 1925.

Avery, Myron. "The Appalachian Trail." *American Forests* 40. 3 (1934): 106-109.

___. "The Skyline Drive and The Appalachian Trail." *PATC Bulletin* (Jan. 1935): 9-10.

___. "Status of the Skyline Drive." *PATC Bulletin* (Jan. 1936): 16-17.

Bancroft, Ernest H. "Why People Should Favor Green Mountain Parkway." *The Vermonter* 41 (January – February 1936): 5-8.

Bayliss, Dudley C. "Planning Our National Park Roads and Our National Parkways." *Traffic Quarterly* 11. 3 (July 1957): 417-440.

Bender, Thomas. "The 'Rural' Cemetery Movement: Urban Travail and the Appeal of Nature." *New England Quarterly* 47. 2 (June 1974): 196-211.

Benson, Harvey P. "The Skyline Drive: A Brief History of a Motorway." *The Regional Review* 4.2 (1940): 3-12.

Berman, Marshall. *All That is Solid Melts into Air: The Experience of Modernity.* 1982. New York: Penguin Books, 1988.

Beveridge, Charles. "Buffalo's Park and Parkway System." *Buffalo Architecture: A Guide.* Reyner Banham, ed. Cambridge, MA: MIT P, 1982.

"Bigger and Better Cities." *The Survey* 54 (1925): 216.

Boyer, M. Christine. *Dreaming the Rational City: The Myth of American City Planning.* Cambridge, MA: MIT P, 1983.

Bronx Parkway Commission. *Annual Report of the Bronx Parkway Commission.* Bronxville: J. B. Lyon, 1906.

___. *Annual Report of the Bronx Parkway Commission.* Bronxville: J. B. Lyon, 1912.

___. *Final Report of the Bronx Parkway Commission.* Bronxville: J.B. Lyon, 1925.

Broome, Harvey. "Origins of the Wilderness Society." *The Living Wilderness* 5.5 (1940): 13-15.

Bryan, Frank. *Yankee Politics in Rural Vermont.* Hanover, NH: U P of New England, 1974.

Bryan, Frank M. and Kenneth Bruno. "Black-topping the Green Mountains: Socio-Economic and Political Correlates of Ecological Decision Making." *Vermont History* 41 (1973): 224-235.

Bryan, Paul T. Introduction. *The New Exploration.* By Benton MacKaye. Urbana: U Illinois P, 1968.

Buder, Stanley. *Visionaries and Planners: The Garden City Movement and the Modern Community.* New York; Oxford: Oxford UP, 1990.

Carr, Ethan. "The Parkway in New York City." *Parkways: Past, Present, and Future.* Boone, NC: Appalachian Consortium P, 1987. 121-128.

___. *Wilderness By Design: Landscape Architecture and the National Park Service.* Lincoln and London: U of Nebraska P, 1998.

Caro, Robert A. "The City-Shaper." *New Yorker* (January 5, 1998): 38-55.

___. *The Power Broker: Robert Moses and the Fall of New York.* New York: Vintage Books, 1974.

Carson, William E. *Conservation and Development in Virginia: Outline of the Work of the Virgnia State Commission on Conservation and Development.* Richmond, VA: Division of Purchase and Printing, 1934.

___. *Conserving and Developing Virginia, Report of W.E. Carson, Chairman, State Commission on Conservation and Development, July 26, 1926 to December 31, 1934.* Richmond: Division of Purchasing and Printing, 1935.

Chadwick, George F. *The Park and the Town: Public Landscape in the 19th and 20th Centuries.* New York and Washington: Frederick A. Praeger, 1966.

Clarke, Gilmore D. "The Parkway Idea." *The Highway and the Landscape.* Ed. W. Brewster Snow. New Brunswick, NJ: Rutgers U P, 1959.

Committee on the Regional Plan of New York and Its Environs. *The Graphic Regional Plan.* Vol. 1. New York: Committee on the Regional Plan of New York and Its Environs, 1929.

___. *The Building of the City.* Vol. 2. New York: Committee on the Regional Plan of New York and Its Environs, 1929.

___. *Regional Survey of New York and Its Environs: Transit and Transportation, and a Study of Port and Industrial Areas and Their Relation to Transportation.* Vol. IV. By Harold M. Lewis, William J. Wilgus, and Daniel M. Turner. New York: Regional Plan of New York and its Environs, 1928.

Cox, Laurie Davidson. "Green Mountain Parkway." *Landscape Architecture* 25.2 (April 1935): 117-126.

___. *Green Mountain Parkway Reconnaissance Survey.* Burlington: Vermont State Chamber of Commerce, 1934.

Cox, Laurie Davidson, Thomas Vint, *et al. The Green Mountain Parkway Final Report by the Landscape Architects of the National Park Service.* Washington, DC: GPO, 1935.

Creese, Walter. *The Search for Environment: The Garden City – Before and After.* Baltimore: Johns Hopkins U P, 1992.

___. *TVA's Public Planning: The Vision, The Reality.* Knoxville: U Tenn. P, 1990.

Dal Co, Francesco. "From Parks to the Region: Progressive Ideology and the Reform of the American City." *The American City: From the Civil War to the New Deal.* Giorgio Ciucci, Francesco Dal Co, Mario Manieri-Elia, and Manfredo Tafuri. Translated from the Italian by Barbara Luigia La Penta. Cambridge, MA: MIT P, 1979. 143-292.

Davidson, Arthur. "Skyline Drive and How it Came to Virginia." *Conserving and Developing Virginia, Report of W.E. Carson, Chairman, State Commission on Conservation and Development, July 26, 1926 to December 31, 1934.* Richmond: Division of Purchasing and Printing, 1935. 75-79.

Demaray, A.E. "Federal Parkways." *American Planning and Civic Annual* 7 (1936): 105-110.

Dench, E.A. "Hiking and Camping Forum." *Nature Magazine* 26 (1936): 186-188.

Downer, Jay. "The Bronx River Parkway." *The Proceedings of the Ninth National Conference on City Planning.* Kansas City: n.p., 1917.

___. "How Westchester County, New York, Made Its Park System." *The Proceedings of the Twentieth National Conference on City Planning.* Philadelphia: Wm. F. Fell, 1928.

Downing, Andrew Jackson. *Rural Essays.* 1853. Ed. George William Curtis. New York: Da Capo P, 1974.

Duffus, R.L. *Mastering a Metropolis: Planning the Future of the New York Region.* New York: Harper & Brothers, 1930.

Eliot, Charles William. *Charles Eliot: Landscape Architect.* 1902. Freeport, New York: Books for Libraries P, 1971.

Eliot, Charles W. II. "The Influence of the Automobile on the Design of Park Roads." *Landscape Architecture Magazine* 13 (1922): 27-36.

Fishman, Robert. "The Regional Plan and the Transformation of the Industrial Metropolis." *The Landscape of Modernity: Essays on New York City, 1900–1940.* Eds. David Ward and Olivier Zunz. New York: Russel Sage Foundation, 1992.

___. *Urban Utopias in the Twentieth Century: Ebenezer Howard, Frank Lloyd Wright, and Le Corbusier.* New York: Basic Books, 1977.

Foglesong, Richard. *Planning The Capitalist City: The Colonial Era to the 1920s.* Princeton: Princeton U P, 1986.

Foresta, Ronald. "Transformation of the Appalachian Trail." *Geographical Review* 77 (1987): 76-85.

Geddes, Patrick. *Cities in Evolution: An Introduction to the Town Planning Movement and the Study of Civics.* 1915. New York: Harper Torchbooks, 1968.

___. "Cities, and the Soils They Grow From." *The Survey Graphic* 44 (April 1, 1925): 40-44.

___. "The First of the Talks from the Outlook Tower: A Schoolboy's Bag and a City's Pageant." *The Survey Graphic* 53 (1925): 525-529.

___. "The Valley in the Town." *The Survey Graphic* 54 (1925): 396-400.

___. "The Valley Plan of Civilization." *The Survey Graphic* 54 (1925): 189-192.

Goldberger, Paul. "Now Arriving: The Restoration of Grand Central Terminal is a Triumphant Validation of an Ambitious Urban Idea." *New Yorker* (September 28, 1998): 92-94.

Goldman, Hal. "James Taylor's Progressive Vision: The Green Mountain Parkway." *Vermont History* 63 (1995): 158-179.

Graves, Henry S. "Road Building in the National Forests." *The American City* 15.1 (1916): 4-10.

Hall, Peter. *Cities of Tomorrow.* Oxford: Blackwell Publishers, 1988.

Harrison, Sarah Georgia. "The Skyline Drive: A Western Park Road in the East." *Parkways: Past, Present, and Future*. Boone, NC: Appalachian Consortium P, 1987. 38-48.

Hill, David P. "Appalachian Heroes as an Indicator of Appalachian Space: Changes in the Meaning of Appalachian Space and Time, 1958-1985." *Parkways: Past, Present, and Future*. Boone, NC: Appalachian Consortium P, 1987. 100-116.

Horan, John F., Jr. "Will Carson and the Virginia Conservation Commission, 1926-1934." *The Virginia Magazine of History and Biography* 92.4 (1984): 391-415.

Hoyle, Raymond J., and Laurie Davidson Cox, eds. *The New York State College of Forestry at Syracuse University: A History of its First Twenty-Five Years, 1911–1936*. Syracuse, NY: The New York State College of Forestry, 1936.

Howard, Ebenezer. 1902 *Garden Cities of Tomorrow*. Cambridge: MIT P, 1965.

Howe, Frederick C. "The Garden City Movement." *Scribner's Magazine* (July 1912): 1-19.

Hughes, Thomas P. and Agatha C. Hughes, Eds. *Lewis Mumford: Public Intellectual*. New York: Oxford U P, 1990.

Ickes, Harold P. "Wilderness and Skyline Drives." *The Living Wilderness* 1.1 (1935): 12.

Jackson, Kenneth T. *Crabgrass Frontier: The Suburbanization of the United States*. Oxford: Oxford U P, 1985.

Johnson, David A. *Planning the Great Metropolis: The 1929 Regional Plan of New York and Its Environs*. London: E & FN Spon, 1996.

___. "Regional Planning for the Great Metropolis." *Two Centuries of American Planning*. Ed. Daniel Schaffer. Baltimore: Johns Hopkins U P, 1987.

Judd, Richard Munson. *The New Deal in Vermont: Its Impact and Aftermath*. New York and London: Garland Publishing, Inc., 1979.

Karsten, Peter, and John Modell. *Theory, Method, and Practice in Social and Cultural History*. New York and London: New York U P, 1992.

Kyle, Robert. "The Dark Side of Skyline Drive." *Washington Post* 17 October 1993: C1.

Lambert, Darwin. *The Undying Past of the Shenandoah National Park*. Boulder, CO: Roberts Rinehart, Inc. 1989.

Lassiter, J.R. "Shenandoah National Park." *The Commonwealth* Document No. 34 (July 1936): 1-8.

The Long Trail News. November 1933 – September 1934.

Lubove, Roy. *Community Planning in the 1920's*. Pittsburgh: U of Pittsburgh P, 1963.

MacKaye, Benton. *The New Exploration*. New York: Harcourt, Brace, and Co., 1928.

___. "An Appalachian Trail: A Project in Regional Planning." *Journal of the American Institute of Architects* 9 (1921): 325-330.

___. "The Appalachian Trail: A Guide to the Study of Nature." *Scientific Monthly* 34 (1932): 330-342.

___. *Employment and Natural Resources*. Washington, DC: Government Printing Office, 1919.

___. "Flankline vs. Skyline." *Appalachia* 20 (1934): 104-108.

___. "The Great Appalachian Trail From New Hampshire to the Carolinas." *New York Times* 18 February 1923, sec. 7: 15.

___. "New York a National Peril." *The Saturday Review of Literature* 23 August, 1930, 68.

___. "The Townless Highway." *The New Republic* 62 (March 12, 1930): 93-95.

___. *From Geography to Geotechnics*. Ed. Paul Bryant. Urbana - Chicago: U Illinois P, 1969.

___. "Progress Toward the Appalachian Trail." *Appalachia* 15 (1922): 244-252.

___. "Some Social Aspects of Forest Management." *Journal of Forestry* 16 (1918): 210-214.

___. "Why the Appalachian Trail." *The Living Wilderness* 1(1935): 7.

MacKaye, Benton, and Lewis Mumford. "Townless Highway for the Motorist: A Proposal for the Automobile Age." *Harper's Magazine* 16 (1931): 347-356.

Marshall, Robert. "The Problem of the Wilderness." *Scientific Monthly* 30 (1930): 141-148.

Mather, Stephen T. "The National Parks on a Business Basis." *The American Review of Reviews* 51 (1915): 429-431.

McClelland, Linda Flint. *Presenting Nature: The Historic Landscape Design of the National Park Service*. Washington, DC: Government Printing Office, 1994.

Mumford, Lewis. *The City in History: Its Origin, Its Transformations, and Its Prospects*. San Diego, New York, London: Harcourt Brace Jovanovich, Publishers, 1961.

___. "Devastated Regions." *The American Mercury* 3 (1924): 217-220.

___. "The Fourth Migration." *The Survey Graphic* 54 (1925): 130-133.

___. *The Golden Day*. New York: Boni and Liveright, 1926.

___. Introduction. *The New Exploration*. By Benton MacKaye. Urbana: U of Illinois P, 1962.

___. *My Work and Days*. New York: Harcourt Brace and Jovanovich, 1979.

___. "The Plan of New York." *The New Republic* 71 (1932): 121-126.

___. "The Plan of New York: II." *The New Republic* 71 (1932): 146-154.

___. "The Plan of New York." (1932) Reprinted in *Planning the Fourth Migration: The Neglected Vision of the Regional Planning Association of America*. Ed. Carl Sussman. Cambridge: MIT P, 1976.

___. "Regions – To Live In." *The Survey Graphic* 54 (1925): 151-152.

___. "Who is Patrick Geddes?" *Survey Graphic* 53 (1925): 523-24.

National Park Service. *The Proposal to Establish the Green Mountain National Park in the State of Vermont*. Washington, DC: The National Park Service, 1937.

New England Regional Planning Commission. *Six for One and One for Six*. Boston: New England Regional Planning Commission, 1937.

Newton, Earle Williams. Preface. *The Role of Transportation in the Development of Vermont*. By William J. Wilgus. Montpelier: Vermont Historical Society, 1945.

Newton, Norman T. *Design on the Land: The Development of Landscape Architecture*. Cambridge: Belknap P of Harvard U P, 1971.

"No Green Mountain Hot-Dogs: Vermonters, Through Town Meetings, Make Sure Barkers Will Not Cry in Their Unspoiled Wilderness by Vetoing Federal Parkway." *The Literary Digest* 121 (1936): 9.

Nolen, John, and Henry V. Hubbard. *Harvard City Planning Studies, Volume XI: Parkways and Land Values*. London: Oxford U P, 1937.

Novack, Frank, ed. *Lewis Mumford & Patrick Geddes: The Correspondence*. London and New York: Rutledge, 1995.

Olmsted, Frederick Law. *The Papers of Frederick Law Olmsted: Writings on Public Parks, Parkways, and Park Systems*. Supplementary Volume I. Eds. Charles E. Beveridge and Carolyn F. Hoffman. Baltimore and London: Johns Hopkins U P, 1997.

Olmsted, Frederick Law, Jr. "Border Roads for Parkways and Parks." *Landscape Architecture* 16 (1926): 67-84.

Olmsted, Frederick Law, Jr., and Theodora Kimball, eds. *Frederick Law Olmsted: Landscape Architect*. New York: The Knickerbocker P, 1928.

Olmsted, John Charles. "Classes of Parkways." *Landscape Architecture* 6 (October 1915 – July 1916): 37-48.

Paris, Louis J. "The Green Mountain Club: Its Purposes and Projects." *Vermonter* 16 (1911): 151-171.

Peach, Arthur W. "Proposed Parkway a Threat to the State's Well Being." *Vermonter* 41 (January – February 1936): 8-13.

Perdue, Charles and Nancy Martin-Perdue. "Appalachian Fables and Facts: A Case Study of the National Park Removals." *Appalachian Journal* 7 (1979-80): 84-104.

___. "'To Build a Wall Around These Mountains': The Displaced People of Shenandoah." *The Magazine of Virginia History* 49 (1991): 48-71.

Radde, Bruce. *The Merritt Parkway.* New Haven and London: Yale U P, 1993.

Regional Plan Association. *From Plan to Reality.* New York: Regional Plan Association, 1933.

Regional Plan Association, Robert Yaro, and Tony Hiss. *A Region at Risk: The Third Regional Plan for the New York-New Jersey-Connecticut Metropolitan Area.* Washington, DC: Island P, 1996.

Reps, John W. *The Making of Urban America: A History of City Planning in the United States.* Princeton: Princeton U P, 1965.

Roper, Laura Wood. *FLO: A Biography of Frederick Law Olmsted.* Baltimore and London: Johns Hopkins U P, 1973.

Rosenzweig, Roy and Elizabeth Blackmar. *The Park and the People: A History of Central Park.* Ithaca: Cornell U P, 1992.

Schaffer, Daniel. "Benton MacKaye: The TVA Years." *Planning Perspectives* 5 (1990): 5-21.

___. *Garden Cities for America: The Radburn Experience.* Philadelphia: Temple UP, 1982.

Schaffer, Daniel, ed. *Two Centuries of American Planning.* Baltimore: Johns Hopkins U P, 1987.

Schuyler, David. *The New Urban Landscape: The Redefinition of City Form in Nineteenth-Century America.* Baltimore: Johns Hopkins U P, 1986.

Seitz, Don C. "Shenandoah National Park: Half a Million Blue Ridge Acres Soon to be Made Public Property." *The Outlook* 144 (1926): 80-83.

Sexton, Roy Lyman Sexton. "The Forgotten People of the Shenandoah." *American Civic Annual* 11 (1930): 19-22.

Shaffer, Marguerite S. "Negotiating National Identity: Western Tourism and 'See America First.'" *Reopening the American West.* Ed. Hal K. Rothman Tucson: U of Arizona P, 1998.

___. "'See America First': Re-Envisioning Nation and Region Through Western Tourism." *Pacific Historical Review* 65 (1996): 559-581.

Shenandoah National Park Association, Incorporated. *Shenandoah: A National Park Near the Nation's Capital.* Richmond, VA: The William Byrd P, Inc., n.d.

"Shenandoah National Park Named By Committee." *National Parks Bulletin* 42 (1924): 1-7.

Showalter, William J. Forward. *Conservation and Development in Virginia: Outline of the Work of the Virginia State Commission on Conservation and Development.* By William E. Carson. Richmond, VA: Division of Purchase and Printing, 1934.

Shurtleff, Arthur A. "A Visit to the Proposed National Park Areas in the Southern Appalachians." *Landscape Architecture* 16 (1926): 67-73.

Silverstein, Hannah. "No Parking: Vermont Rejects the Green Mountain Parkway." *Vermont History* 63 (1995): 133-157.

Simmons, Dennis E. "Conservation, Cooperation, and Controversy: The Establishment of Shenandoah National Park, 1924-1936." *Virginia Magazine of History and Biography* 89 (1981): 387-404.

Southern Appalachian National Park Commission. *Final Report of the Southern Appalachian National Park Commission to The Secretary of the Interior, June 30, 1931.* Washington, D.C.: GPO, 1931.

Spann, Edward K. *Designing Modern America: The Regional Planning Association of America and Its Members.* Columbus: Ohio State U P, 1996.

Stein, Clarence. Introduction. "An Appalachian Trail: A Project in Regional Planning." By Benton MacKaye. *Journal of the American Institute of Architects* 9 (1921): 325-330.

___. "Dinosaur Cities." *The Survey* 54 (1925): 135-138.

Steltzer, Irwin. "Don't Believe the Doomsayers." *New York Post* 21 March, 1996.

Sussman, Carl. Ed. *Planning the Fourth Migration: The Neglected Vision of the Regional Planning Association of America.* Cambridge: MIT P, 1976.

Swain, Donald C. *Wilderness Defender: Horace M. Albright and Conservation.* Chicago and London: U of Chicago P, 1970.

Thomas, John L. "Lewis Mumford, Benton MacKaye, and the Regional Vision." *Lewis Mumford: Public Intellectual.* Ed. Thomas P. and Agatha C. Hughes. New York: Oxford U P, 1990.

Thoreau, Henry David. *A Week on the Concord and Merrimack Rivers.* 1849. New York: Book of the Month Club, 1996.

___. *The Maine Woods.* 1864. New York: Book of the Month Club, 1996.

___. *Walden.* 1854. New York: Book of the Month Club, 1996.

Upson, William Hazlett. "The Green Mountain Parkway." *Informational Bulletins on State Problems 4.* Burlington: Vermont State Chamber of Commerce, 1934.

The Vermont State Chamber of Commerce. "Col. William J. Wilgus Explains the Proposed Green Mountain Parkway." *Bulletin of the Vermont State Chamber of Commerce* August 28, 1933.

Virginia State Chamber of Commerce. "Shenandoah National Park in Virginia." Document No. 3 (1925). Richmond, VA.

Waterman, Laura, and Guy Waterman. *Forest and Crag: A History of Hiking, Trail Blazing, and Adventure in the Northeast Mountains.* Boston: Appalachian Mountain Club, 1989.

Weigold, M.E. "Pioneering in Parks and Parkways: Westchester County, New York, 1895-1945." *Essays in Public Works History* 9 (1980): 1-43.

Westchester County Park Commission. *Annual Report* (1926-1930). White Plains, NY: The Westchester County Park Commission, 1926-1930.

Wilgus, William J. *The Role of Transportation in the Development of Vermont.* Montpelier: Vermont Historical Society, 1945.

___. "Transportation in the New York Region." *Transit and Transportation, and a Study of Port and Industrial Areas and Their Relation to Transportation.* New York: Regional Plan of New York and its Environs, 1928.

___. "Vermont's Opportunity." *Bulletin of the Vermont State Chamber of Commerce,* August 8, 1933.

Wilhelm, Gene, Jr. "Shenandoah Resettlements." *Pioneer America: The Journal of Historic American Material Culture* 14:1 (March 1982): 15-40.

Zaitzevsky, Cynthia. *Frederick Law Olmsted and the Boston Park System.* Cambridge, MA and London, England: Belknap P of Harvard U P, 1982.

Zapatka, Christian. "The American Parkways: Origins and Evolution of the Park-Road." *Lotus International* 56 (1987): 96-125.

___. *The American Landscape.* Ed. Mirko Zardini. New York: Princeton Architectural P, 1995.

Manuscript Collections and Government Documents

Byrd, Harry F. Executive Papers, 1926-1930. Virginia State Library. Richmond, VA.

Green Mountain Club Archives. Cowles Papers. Green Mountain Parkway, 1933-1937. Vermont Historical Society. Montpelier, VT.

MacKaye Family Papers. Special Collections. Dartmouth College Library. Hanover, NH.

National Park Service. Records on the Shenandoah National Park and the Skyline Drive. Record Group 79. Central Classified Files, 1907-1949, 1907-1932. General Files. National Archives at College Park, MD.

___. Records of the Shenandoah National Park. Record Group 79. Central Classified Files, 1907-1949, 1907-1932. National Parks: Shenandoah. National Archives at College Park, MD.

National Park Service Archive. National Park Service Center Archive. Harpers Ferry, WV.

Shenandoah National Park Papers. Shenandoah National Park Headquarters, Luray, VA.

State Conservation and Development Commission of Virginia. Programs of Meetings. Vol. 1. 1926-1927. Record Group 9/F/9/1/1. Virginia State Library. Richmond, VA.

Taylor, James P. Collected Papers. Vermont Historical Society, Montpelier, VT.

U.S. Forest Service. Records of the U.S. Forest Service. Record Group 95. National Archives at College Park, MD.

Wilderness Society Collection. Denver Public Library. Denver, CO.

Zerkel, Ferdinand L. Collected Papers. Archives, Shenandoah National Park Headquarters, Luray, VA.

Other Non-Published Material

Ellis, Clifford Donald. "Visions of Urban Freeways, 1930-1970." Diss. U California at Berkeley, 1990.

Goldman, Hal. "'Vermont's Opportunity': Responses to the Green Mountain Parkway." Master's thesis, U of Vermont, 1995.

Gutfreund, Owen D. "20th Century Sprawl: Accommodating the Automobile and the Decentralization of the United States." Diss. Columbia U, 1998.

Lambert, Darwin. "Shenandoah National Park, Administrative History, 1924-1976." Unpublished Manuscript, Shenandoah National Park Archives, Luray, VA, 1979.

Mallonee, Walter W. *Origin of the Skyline Drive Through the Shenandoah National Park in the Blue Ridge Mountains of Virginia*. United States: WW Mallonee, 1995.

Mason, Randall F. "Memory Infrastructure: Preservation, 'Improvement' and Landscape in New York City, 1898–1925." Diss. Columbia U, 1999.

Orlin, Glenn S. "The Evolution of the American Urban Parkway." Diss. George Washington U, 1992.

"The Reminiscences of Gilmore David Clarke: Oral History." Oral History Collection, Columbia University, 1959.

Silverstein, Hannah. "A Road Not Traveled: Vermont's Rejection of the Green Mountain Parkway." Senior Thesis, Yale U, 1994.

Simmons, Dennis E. "The Creation of Shenandoah National Park and the Skyline Drive, 1924-1936." Diss. U of Virginia, 1978.

Sutter, Paul S. "Labor and Natural Resources: Colonization, the Appalachian Trail, and the Social Roots of Benton MacKaye's Wilderness Advocacy." Presented at the University of Virginia History Department Seminar. Charlottesville, VA. September 28, 1998. Unpublished manuscript in the possession of the author.

___. "Driven Wild: The Intellectual and Cultural Origins of Wilderness Advocacy During the Interwar Years (Aldo Leopold, Robert Sterling Yard, Benton MacKaye, Bob Marshall)." Diss. U of Kansas, 1997.

Taylor, Virginia. Personal interview conducted by Matthew Dalbey. 25 Feb. 1994.

Whitehead, Joseph. "The Proposed Shenandoah National Park in Virginia." Speech in the United States House of Representatives. Feb. 20, 1926.

Online Sources

"Building the Trail: 1921-1937." *Appalachian Trail Conference: Building the Trail Page.* January 14, 1999. < www.appalachiantrail.org/about/history/index.html>.

The Green Mountain Club Homepage. January 15, 1999. <www.greenmountainclub.org>.

Paris, Louis J. "Purposes and Membership" from *The Long Trail Guidebook* 1921. *The Green Mountain Club Homepage.* January 15, 1999. <www.greenmountainclub.org>.

Government Documents

Journal of the House of the State of Vermont. Biennial Session 1935. "An All-Vermont Plan." Published by Authority, Montpelier, VT, 1935. 232-234.

Index